NH$6/00

MALAYAN NATURE HANDBOOKS

General Editor

M. W. F. TWEEDIE

THE aim of the Malayan Nature Handbooks is to provide a series of handy, well-illustrated guides to the fauna and flora of the Malay Peninsula. They can, of course, be no more than introductory; the animal and plant life of Malaya is on such a lavish scale that comprehensive accounts of the groups described in each of the Handbooks must be either severely technical or voluminous and correspondingly costly. The selection of species described in each one has been carefully made, however, to illustrate those most likely to be the first encountered by reasonably observant people residing in or visiting Malaya; reference to rarities or species confined to inaccessible country has been avoided, except where such species are of special interest.

It is the Editor's belief that interest in animals and plants is best aroused by providing the means of identifying and naming them. The emphasis of the Handbooks is therefore firstly on identification, but as much information on habits and biology is included as space will allow. It is hoped that they may be of use to schools in supplementing courses in nature study and biology, and a source of pleasure to that quite numerous assemblage of Malayans whose complaint has been that they would gladly be naturalists if someone would show them the way.

OTHER TITLES IN THE SERIES

M. W. F. Tweedie COMMON MALAYAN BIRDS

MALAYAN NATURE HANDBOOKS

Common Malayan Butterflies

BY R. MORRELL

ILLUSTRATED BY A. H. BURVILL

LONGMANS

LONGMANS MALAYSIA SDN. BERHAD
44 Jalan Ampang, Kuala Lumpur.

Associated companies, branches and
representatives throughout the world.

© R. Morrell, 1960

First published 1960
*New impression *July 1969*

SBN 582 69447 7

Printed in Hong Kong
by Dai Nippon Printing Co. (International) Ltd.

CONTENTS

LIST OF PLATES

Plate 5 A GREAT MORMON (*Papilio memnon agenor*), ♂; B, ♀-form *distantianus*

6 A FIVE-BAR SWORDTAIL (*Graphium antiphates itamputi*), ♂
B *Graphium evemon*
C MALAYAN JEZEBEL (*Delias ninus*)
D COMMON BLUEBOTTLE (*Graphium sarpedon luctatius*), ♀
E PAINTED JEZEBEL (*Delias hyparete metarete*), ♂ underside; F, ♀

7 A CHOCOLATE ALBATROSS (*Appias lyncida vasava*), ♂; B, ♀
C GREAT ORANGE TIP (*Hebomoia glaucippe aturia*), ♂
D ORANGE ALBATROSS (*Appias nero figulina*), ♂
E WANDERER (*Valeria v. lutescens*), ♂ (♀, see Plate 2 G)
F COMMON GRASS YELLOW (*Eurema hecabe contubernalis*)
G *Eurema sari sodalis*, underside

8 A LEMON MIGRANT (*Catopsilia p. pomona*), ♂; B, ♀
C BLACK-VEINED TIGER (*Danaus melanippus hegesippus*), ♂
D PLAIN TIGER (*Danaus chrysippus alcippoides*), ♂ (form *chrysippus*)
E COMMON TREE NYMPH (*Idea jasonia logani*), ♂

9 A STRIPED BLUE CROW (*Euploea m. mulciber*), ♂; B, ♀
C BLUE-BRANDED KING CROW (*Euploea leucostictos leucogonis*), ♂
D STRIPED BLACK CROW (*Euploea eyndhovii gardineri*), ♂

10 A LARGE CHOCOLATE TIGER (*Danaus sita ethologa*), ♂
B COMMON THREE-RING (*Ypthima pandocus corticaria*), underside
C COMMON FOUR-RING (*Ypthima ceylonica huebneri*)
D COMMON EVENING BROWN (*Melanitis l. leda*), upperside; E and J, undersides
F MALAYAN BUSH BROWN (*Mycalesis f. fuscum*); G, underside
H NIGGER (*Orsotriaena medus cinerea*), underside
I BAMBOO TREE BROWN (*Lethe europa malaya*), underside

Plate 11 A COMMON FAUN (*Faunis canens arcesilas*)

B YELLOW-BARRED (*Xanthotaenia b. busiris*)

C COMMON PALMFLY (*Elymnias hypermnestra beatrice*), ♂; D, subsp. *agina* ♀; G, subsp. *tinctoria* ♀

E SATURN (*Zeuxidia a. amethystus*), ♂; F, ♀

12 A PALM KING (*Amathusia phidippus chersias*), ♂; B, ♀; C, underside

D DARK BLUE JUNGLE GLORY (*Thaumantis klugius lucipor*), ♂

13 A RUSTIC (*Cupha erymanthis lotis*)

B VAGRANT (*Vagrans egista macromalayana*)

C BANDED YEOMAN (*Cirrochroa o. orissa*)

D ROYAL ASSYRIAN (*Terinos terpander robertsia*)

E CRUISER (*Vindula arsinoe erotella*), ♂; F, ♀

G BLUE PANSY (*Precis orithya wallacei*), ♂

H PEACOCK PANSY (*Precis almana javana*), ♀

14 A MALAY LACEWING (*Cethosia hypsea hypsina*), ♂; D, underside

B GREAT EGG-FLY (*Hypolimnas b. bolina*), ♂; C, ♀

E AUTUMN LEAF (*Doleschallia bisaltide pratipa*)

F STRAIGHT-LINE MAPWING (*Cyrestis nivea nivalis*)

15 A BURMESE LANCER (*Neptis heliodore dorelia*)

B COMMON SAILOR (*Neptis hylas marmaja*), upperside; C, underside

D LANCE SERGEANT (*Parathyma pravara helma*)

E COMMANDER (*Moduza procris milonia*)

F CLIPPER (*Parthenos sylvia lilacinus*)

G COLOUR SERGEANT (*Parathyma nefte subrata*), ♂; H, ♀-form *neftina*; I, ♀-form *subrata*

Plate 16 A ARCHDUKE (*Euthalia dirtea dirteana*), ♂; B, ♀

C COMMON NAWAB (*Polyura athamas samatha*)

D MALAY VISCOUNT (*Tanaecia p. pelea*), ♀

E BARON (*Euthalia aconthea gurda*), ♂

F HORSFIELD'S BARON (*Euthalia iapis puseda*), ♂; G, ♀

17 A TAWNY RAJAH (*Charaxes polyxena*)

B SMALL HARLEQUIN (*Taxila t. thuisto*), ♂; C, ♀

D PUNCHINELLO (*Zemeros flegyas albipunctata*), ♀

E LESSER DARKIE (*Allotinus unicolor*), ♂; F, ♀ underside

G BIGGS'S BROWNIE (*Miletus b. biggsii*)

H COMMON LINE BLUE (*Nacaduba nora superdates*)

I SUMATRAN GEM (*Poritia s. sumatrae*), ♂; J, ♀

K MALAYAN PLUM JUDY (*Abisara saturata kausambioides*), ♂; L, ♀

M MALAY RED HARLEQUIN (*Laxita damajanti*), ♂

18 A COMMON HEDGE BLUE (*Celastrina puspa lambi*), upperside; B, underside

C LESSER GRASS BLUE (*Zizina otis lampa*), ♂; D, underside

E PEA BLUE (*Lampides boeticus*), ♀ upperside; F, underside

G STRAIGHT PIERROT (*Castalius roxus pothus*), underside

H SKY BLUE (*Jamides c. caeruleus*), ♂; I, ♀ underside

J YAM FLY (*Loxura atymnus fuconius*)

K COMMON POSY (*Marmessus ravindra moorei*), ♂; L, underside

M PEACOCK ROYAL (*Pratapa cippus maxentius*), ♂; N, ♀

O MALAYAN SUNBEAM (*Curetis santana malayica*), ♂; P, underside

Q BLACK-BANDED SILVERLINE (*Spindasis syama terana*), underside

Plate 19 A CENTAUR OAK BLUE (*Narathura c. centaurus*), ♂; J, underside

B *Aurea trogon*, ♂; C, ♀

D SMALL TAIL-LESS OAK BLUE (*Narathura antimuta*), ♂; E, ♀

F *Rapala dieneces*, ♂; G, ♀; H, underside

I COMMON RED FLASH (*Rapala i. iarbus*), ♂

K FLUFFY TIT (*Zeltus amasa maximinianus*), ♂; L, underside

M BRANDED IMPERIAL (*Eooxylides tharis*), underside

20 A HOARY PALMER (*Unkana ambasa batara*)

B ORANGE AWLET (*Bibasis harisa consobrina*), ♂; C, ♀

D PLAIN BANDED AWL (*Hasora v. vitta*), underside

E LARGE SNOW FLAT (*Tagiades g. gana*)

F BUSH HOPPER (*Ampittia dioscorides camertes*), ♂; H, ♀

G PALM DART (*Telicota a. augias*), ♂

I LESSER DART (*Potanthus o. omaha*), ♂

J STARRY BOB (*Iambrix stellifer*), underside

K YELLOW-VEIN LANCER (*Plastingia l. latoia*)

L SMALL BRANDED SWIFT (*Pelopidas m. mathias*)

M BANANA SKIPPER (*Erionota t. thrax*)

N DARK YELLOW-BANDED FLAT (*Celaenorrhinus aurivittata cameroni*)

O GRASS DEMON (*Udaspes folus*)

P COCONUT SKIPPER (*Hidari iravi*), underside

Q CHESTNUT BOB (*Iambrix s. salsala*)

R CONTIGUOUS SWIFT (*Polytremis l. lubricans*)

To

S. C.-P., P. F.-S. and G. R. M.

ACKNOWLEDGMENTS

I should like to thank all those who have helped directly or indirectly with the production of this book, and particularly Dr. R. Hartland-Rowe of Makerere College for reading the proofs, and Mr. T. G. Howarth and the Trustees of the British Museum for providing facilities for the artist to prepare illustrations of such species as were not in my own collection.

R. M.

To these acknowledgments I should like to add a note of my debt to Lt. Col. C. F. Cowan for checking the names for this, the second, edition, and my thanks to him for the corrections which follow.

CORRECTIONS AND CHANGES IN NOMENCLATURE

Page 42	For '*Neptis hylas marmaja*' read '*Neptis hylas mamaja*'.
Pages 52–53	The late Brigadier W.H. Evans's division of the genus *Arhopala* does not seem to have been generally adopted (see J.N. Eliot, *Malayan Nature Journal*, Vol. 17 pp. 188–217). So for '*Narathura centaurus*' read '*Arhopala centaurus*'; for '*N. aedias agnis*' and other members of this genus, read '*A. aedias agnis*' etc., and for '*Aurea trogon*', read '*Arhopala trogon*'. This last correction also applies to the caption for Plate 19 which is printed at the foot of page 55. Line 10 of page 53 should now read: 'distinguished from *A. eumolphus maxwelli* by their' ....
Page 54	For '*Marmessus ravindra moorei*' read '*Drupadia ravindra moorei*'.
Page 57	For '*Hidari iravi*' read '*Hidari irava*'.

INTRODUCTION

THIS book is intended to be a brief introduction to a fascinating and virtually inexhaustible subject. Of the nine hundred butterfly species recorded from Malaya, we can describe only a selection, concentrating upon about a hundred species which the collector is most likely to encounter on his earlier rambles, and giving some hint of the treasures which await him when he ventures farther into the jungle. A few species, so beautiful or otherwise remarkable that they are famous the world over, have been included although they are not very common; but for the most part the commoner species have been preferred.

The reader will find more detailed and extensive information in the magnificent *Butterflies of the Malay Peninsula* by Corbet and Pendlebury, a recently published and most careful compilation of all the significant facts known when it was written. It contains anatomical notes and diagrams, maps of distribution, and detailed keys for tracing the species of every genus. Every serious student will have need to refer to this book sooner or later, and for many years it must remain the standard work on the subject. It is possible, however, that the beginner may be discouraged by the very qualities which will render the book invaluable to him later, and this small volume is designed to give him the information he requires initially, and to awaken further interest.

With reasonable care, most butterflies retain their beautiful colours almost indefinitely, and this is one reason that they appeal to collectors. Fortunately, the mere collector, for whom a butterfly's life-history 'begins in the net and ends in the killing bottle', often turns into a more serious observer. For such people there is plenty of scope in Malaya, for much remains to be discovered about the habits and early stages even of many common butterflies.

And as soon as the observer turns his attention to life-histories, he will find that the study of butterflies is no narrow interest, but provides an introduction to much of the Malayan natural scene as a whole. He acquires a knowledge of the flowers that the butterflies visit, and the plants their larvae feed on; observing their protective devices, he finds himself considering the habits of their enemies, the insectivorous birds and reptiles, and the other insects which prey upon butterflies and butterfly larvae, or live with these in symbiotic relationship; his eyes will be sharpened to notice many things of which he was unaware, or but dimly aware, earlier.

The student of our butterflies will indeed catch glimpses of what may, without inaccuracy, be called the meaning of life in the tropics: the incredible ingenuity and efficiency by which all the innumerable animal species, large and small, each jostling for breathing space, manage to survive in a competition so intense and continuous that it has many exciting manifestations.

Collecting. The art of collecting can be learnt only by experience. Basic information, by the lack of which the beginner might be handicapped, is given below: valuable additional information is given in one of the Malayan Museum Pamphlets, *Collecting Butterflies in Malaya*, by M. W. F. Tweedie, formerly Director of Raffles Museum.

Nets should be strong, large and light. Some dealers[1] make a serviceable 'large kite net' frame of lightweight metal, but an even larger one can be made of rotan. At any rotan shop select a long piece, ten or eleven feet long if possible, and about five-eighths of an inch in diameter. Shape it carefully over a flame, and bind it with wire as shown in the figure. The bag can be made of any dark green or black netting (several excellent types of net are made nowadays, lighter and stronger than ordinary mosquito netting), topped with three inches of stout cloth of the same colour to take the wear of the cane; and the bottom, the closed end, must be rounded. The whole frame is about four feet six inches long and provides a fair reach, but is not too big to be used effectively along jungle paths where movements are restricted. It is, moreover, so light that it can easily be bound to a longer stick should this become necessary.

Unless the insect is on the ground, use an upwards or sideways sweep of the net with a 'follow through' in which the frame is twisted round, making a pocket with the insect inside. If it is a large one it

[1] Entomological equipment may be obtained from Messrs. Watkins & Doncaster, 110 Parkview Road, Welling, Kent, U.K.

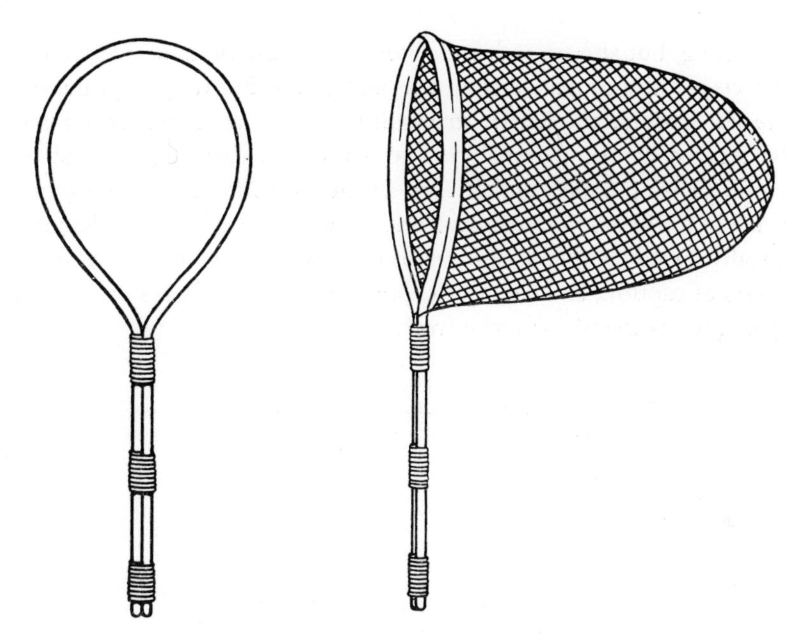

Fig. 1 The construction of a butterfly net with rotan frame

may be stunned by pinching the thorax below; it may then be papered, and then, still in paper, placed in the killing bottle. Small butterflies should go direct into the bottle, then, if *rigor* reverses the wings, these should be blown back into position, the body being gripped with forceps, and the insect then papered. Most Chinese village shops stock cheap forceps or 'tweezers' at ten or twenty cents for entomologists who realize, on their way to the hunting ground, that these instruments have been left at home.

A chemist who knows you well will make up a cyanide killing bottle if you supply him with a well-stoppered jar. You may make a less dangerous one yourself by merely putting a wad of cotton-wool at the bottom of a jar, soaking it with chloroform or ethyl acetate, and putting discs of clean blotting-paper on top.

The papers which have been mentioned are rectangular pieces of stiffish paper folded as indicated in the diagram (the line *XY* is the first fold, so that *A* comes on to *B*). These papers are useful not only for protecting butterflies from damage in bottle or tin during a day's collecting, but also for indefinite storage in some container where the insects are also protected from ants and damp. But the longer insects are left like this the harder they will be to set; larger butterflies can be relaxed by being transferred to a tin with clean damp sand or damp blotting-paper, and left for twenty-four hours or longer, but small butterflies become difficult and sometimes impossible to handle. Ideally, butterflies should be put into a relaxing tin within twenty-four hours of capture, and set after another twenty-four hours or as soon thereafter as the wings move freely.

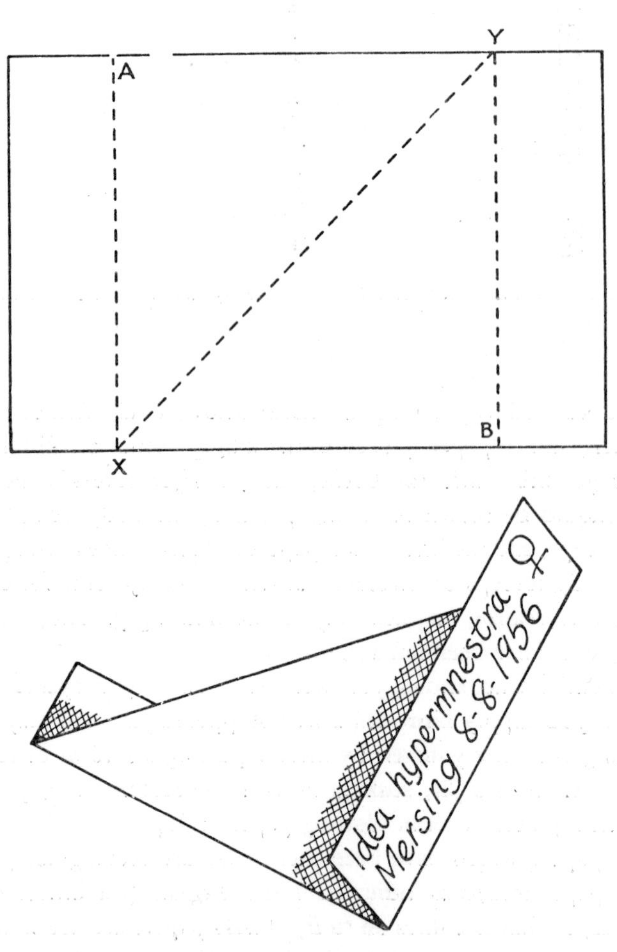

Fig. 2 Folded paper envelope for butterflies

'Setting' is the process of arranging the wings in the position they are required to retain permanently, and leaving them to dry on a 'setting board'. First the butterfly should be transfixed by an entomological pin (ordinary pins are too thick and blunt) passing through the thorax above. Care should be taken to get the pin straight and the body the right height in the groove. With a long pin or setting needle pressing against the wing veins, the wings are flattened and moved forward, and held with narrow paper strips or braces until they can be secured with additional braces, or until the braces can be replaced by broader strips of transparent paper. It is not easy to set well, but skill will come with practice. Uniformity must be the aim: not only is a badly set collection unpleasant to look at, but differences

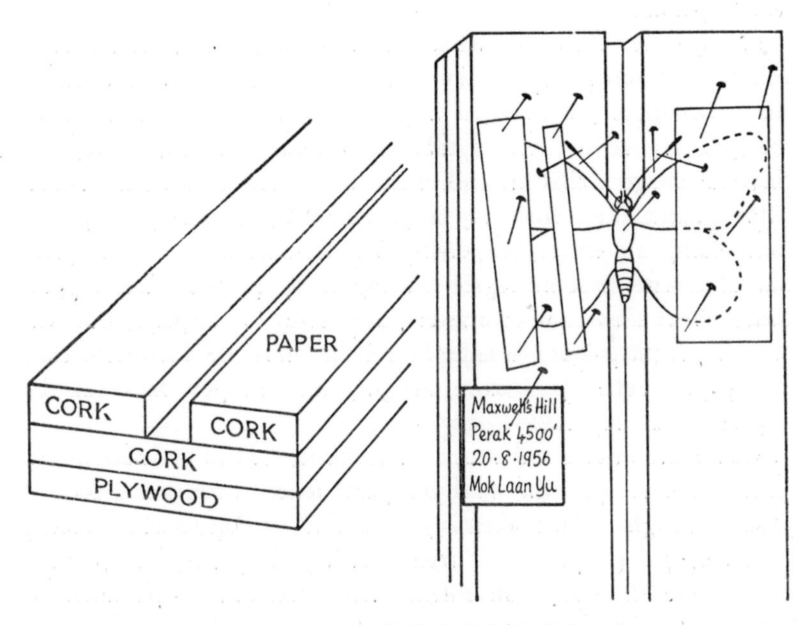

Fig. 3 *Left,* diagram of a setting board; *right,* a butterfly on the setting board, to show method of setting

between closely related species, or significant variations within one species, can be seen only if all the insects are set uniformly. The standard position is with the rear edge of the forewing (the forewing 'dorsal margin') at right angles to the body, and the hindwing pushed forward until only a little is hidden below the forewing, as in the diagram. Body and antennae should be straightened, and insects should be examined after twenty-four hours to make sure that nothing has slipped or worked loose. At most times of the year in Malaya, boards are best put into a drying drawer with a 15- or 25-watt lamp to hasten the process, but it will also be necessary to disinfect the drawer thoroughly with DDT and paradichlorbenzene.

Every insect must be labelled with (*a*) place of capture, (*b*) the date, and (*c*) any other information deemed necessary, e.g. collector's name, altitude of capture, food-plant (if the insect has been bred). If all this is written on a small square of stiff paper and pinned beside the insect on the board, it can be transferred to the insect's own pin (beneath the body) as soon as the butterfly is set. Thus no mistake will be possible.

Finally, after about two weeks on the setting board, the butterfly will be dry enough to be removed to its place in the collection. Butterflies should be arranged in systematic order in storeboxes or in the glass-topped drawers of cabinets. Cabinets are expensive, but their advantage is that the collection can be seen as often as desired without harm. A storebox may be quite airtight, but when it is opened fresh damp air enters, dispersing disinfectants and, in a tropical coastal climate, to some degree relaxing the specimens. Glass-topped boxes, other than cabinet drawers, are useful for displays, but will in time permit colours to fade. Storeboxes must be lined with cork and paper: better (i.e. more airtight) and cheaper ones can be imported than any made locally.

Mould and insects will attack the collection unless precautions are taken. Creosote (not the kind you paint fences with, but a refined product bought from a chemist) on a wad of cotton-wool erected on a long pin (to prevent it from staining the paper) will prevent mould, and the most convenient insect deterrent is paradichlorbenzene. The latter can be put into the little cell made for the purpose in imported storeboxes, the paper between cell and box being pin-pricked; or a small bag of it may be pinned alongside the creosote wad.

Paths, clearings and streams in the jungle are the best collecting grounds. Some species are to be found congregated at moist spots on forest roads, and on sandbanks by rivers, where animals have come down to drink and have contaminated the sand with urine. Still more species frequent mountain-tops; and the tops of even small hills are worth visiting.

Some trees and shrubs are particularly attractive to butterflies of all families, and the collector will soon learn to recognize these and look out for them.

Some Satyrids, most Amathusiids and a few Nymphalids are attracted to rotting papaya, banana or, best of all, pineapple. The fruit can be nailed to a tree, beneath a piece of wire gauze to protect it from monkeys and squirrels, or stamped into the ground, or hung from a string. Some are attracted to carrion and decaying fish or prawns, or to dung: tiger and panther dung are said to be the best. A seemingly magically effective bait for all Danaids (but for no other butterflies in Malaya) is the partly dried plant of *Heliotropium indicum*. A bunch of ten or twelve plants pulled up by the roots and hung up to dry will start to draw Danaids within forty-eight hours; after four days the plants should be well sprinkled once a day if rain does not render this unnecessary. If the spot is a suitable one, fairly exposed but not directly sunlit, and in a district where Danaids are plentiful, the butterflies will continue to come for a fortnight.

Despite the splendour and variety of forest insects, the collector may wish to commence operations with the butterflies of the gardens and roadsides near his home. This preference is commendable, since only here can observations be continuous and more than merely superficial. It is here that he will best appreciate the butterfly's life as part of a bigger scene, observe its predators, its seasonal changes, the fluctuations in its population, the influence upon it of the weather and of the seasonal changes of plants; it is here that he will see butterflies court, mate and lay eggs, and doubtless it will be butterflies of gardens and open country that he will first breed.

Even in this field he may contribute to our knowledge, for several of these common butterflies have not been bred hitherto, so superficial have observations been. Often a 'life-history' is limited to a note of the colour of the larva, the name of the food-plant and, perhaps, of the time spent as larva and pupa; the last a meaningless record because it might well be different under natural conditions, or vary according to the season of the year, or the latitude, or the altitude. There is scope, too, for experiments in genetics, although a basic difficulty is that of getting butterflies to mate in captivity and thus breeding from strains of which the genetic constitution is known. Much of our ignorance will not be dispersed until butterfly breeding houses are made as large as those which Dr. William Beebe constructed in Trinidad—large enough for the insects to fly and bask and mate in, to visit their favourite flowers, and lay their eggs on the larval food-plants.

But for at least some experiments a small breeding cage will do, and if the prime concern is simply to obtain the adult, the main thing to remember is that the moisture, of food-plant and of the larva itself, must be conserved. Young larvae feeding on leaves only in bud should be kept in glass tubes, small jars or the small glass-topped tins which dealers sell for this purpose. Later, sprigs of bigger leaves can be placed with their stems in a small bottle of water (e.g. an ink bottle), and the whole placed in a larger and more airy cage. But moisture is still important; most larvae drink as well as eat, and it is advisable to sprinkle water on the leaves from time to time. On the other hand caterpillars can drown and, unless cotton-wool is wrapped round the stems in the neck of the bottle, many larvae will walk straight down the stem into the water. Never handle larvae, particularly when they may be about to moult or to pupate. The best way of transferring them to new food is to prepare a fresh supply, and then detach from the old foliage any leaves or parts of leaves on which larvae are resting, and pin these bits of old foliage on to the new.

From three days to four weeks after the larva has pupated, the colours of the butterfly's forewing will become visible through the pupal case, and the next day it will emerge. With a few exceptions (some of the Skippers, for instance) butterflies emerge in the morning, but an hour or so is spent in inflating, drying and easing their wings before a trial flight. They should not be killed before this process has been completed.

A breeding cage of the type B illustrated at Fig. 4 can be bought ready-made at many furniture shops, where they are sold for the purpose of protecting food from flies. When the middle shelf is removed it is serviceable, but not ideal; it is not high enough and the perforated zinc in the door, side and back panels prevents clear observation. Type A will have to be made specially but will prove

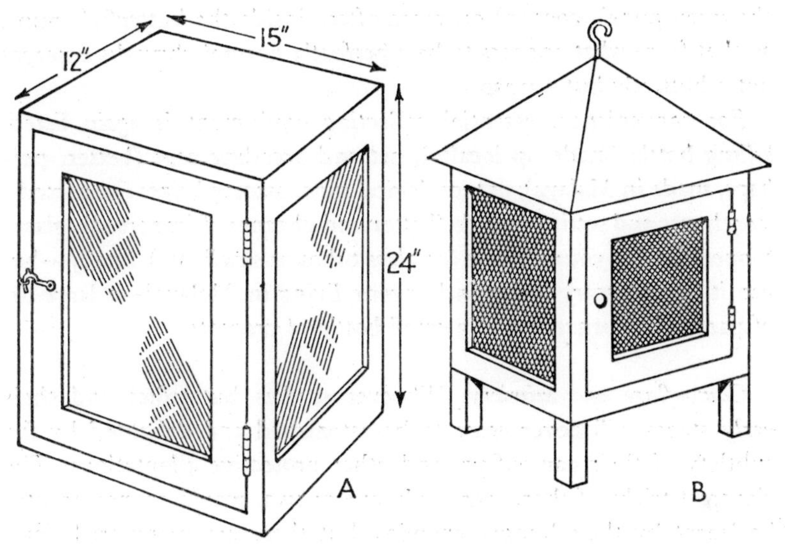

Fig. 4. Two types of breeding case.

more satisfactory. The panel in the door is of glass, and so are the side panels, while that at the back is of perforated zinc.

One method of breeding is to 'sleeve' the food-plant by putting a bag of mosquito netting on the growing plant over the spray of leaves on which the larvae are feeding, the open end being tied round the branch or stem. Sometimes this is the best method: for instance, if the plant is of a kind which withers when cut and put in water. It also appears that female butterflies will lay eggs on the growing plant, if imprisoned in a sleeve, more readily than on cut sprays in a breeding cage.

If breeding provides excitements and surprises, it brings disappointments too: a beginner's failures are often due to forgetfulness or carelessness over the provision of fresh leaves for his larvae. But sometimes caterpillars die for no apparent reason, and often for reasons only too apparent, but beyond the control of the entomologist. A caterpillar may fail to develop normally because it is parasitized; it at last ceases to feed and a host of tiny grubs eat their way out through its skin and spin little cocoons over its body; in due course small flies emerge. Often the parasite is a single wasp grub, but this pupates either inside the caterpillar (which dies and shrivels round the wasp pupa's cocoon) or, more often, inside the butterfly's pupa, so that from what appears to be a perfectly normal chrysalis emerges not a butterfly but a wasp.

For convenience, essential collecting equipment is again listed: killing bottle (made up locally); net and breeding cage (better, perhaps, made in Malaya); entomological pins, assorted sizes (imported); storeboxes and setting boards (better and cheaper if imported); glass-topped tins for young larvae (or glass tubes or small jars); finally—but can it be necessary to remind anyone living in Malaya?—a large tin of paradichlorbenzene and a small bottle of creosote.

Camouflage and mimicry. Whoever studies butterflies and their early stages will never cease to be astonished and delighted by the subtlety of their camouflage and other protective adaptations. The cleverest tricks, if there were only one or two, would sooner or later be learnt by their hungry enemies; but there are many tricks and hundreds of variations, and it seems as though every loophole of possible survival, every device which might give an advantage, however slight and temporary, is being tried out. A caterpillar may resemble a curled leaf, a leaf vein, a bird-dropping, a bud, a stalk end, even the head and neck of a snake; a live pupa may be disguised as a leaf, growing or withered, an unripe berry, a broken-off twig, or a pupa case already empty. Many larvae habitually rest crookedly, so that their symmetry is broken and they are unrecognizable; some build themselves hides or barriers or backgrounds against which their camouflage is more effective; others construct decoys—inedible dummy caterpillars to distract attention from themselves. Often the 'flash coloration' of the upperside of a butterfly in flight deludes the pursuer into searching for something quite different in appearance from the insect settled with only its sombre underside showing. The butterfly most famous for protective adaptation is the Leaf Butterfly (page 41), but although no other produces so neat a disguise for display in a museum showcase, there are many just as effective in natural surroundings—in undergrowth or on a leaf-strewn forest floor.

Centuries before humans understood the simplest principles of camouflage, these insects were employing the most elaborate. Not, of course, with conscious purpose; but chance improvements had given their owners a decisive advantage in the struggle for life. Only the best-protected forms were able to reach maturity and so breed and pass on their improved forms, and tendency to improve, to another generation. A female butterfly may lay up to two hundred eggs, and to maintain the population only two of these potential lives need reach maturity. This will indicate how much wastage there is, and how much material for selection to act upon.

Even the eggs are in danger from creatures which will devour them outright or inject into them their own minute eggs that will hatch into tiny parasitic grubs. For young larvae too the main danger is probably from other insects, parasitic flies and wasps; later lizards and birds take their toll—one pair of birds with young nestlings may take them hundreds of caterpillars a day. Once a caterpillar has pupated the insect may be reasonably safe; most pupae are well camouflaged, and in any case they have the great advantage of remaining motionless. At last the mature insect emerges and can mate.

But even maturity is a comparative thing; not so much for the male, whose job is soon done, but for the female. With mating, her work has only begun. For the species, it clearly matters a great deal whether she is eaten today, after laying ten eggs, or at the end of next week, after laying two hundred. And while she is laying eggs, seeking out the right food-plant by smell, making sure she is not laying on some interlacing creeper, choosing the young shoots, and generally sticking at her job as if she were prompted by conscience or maternal solicitude instead of just instinct, she is especially vulnerable. Hence arises the fact that the female is often quite different from the male, and usually far less conspicuous—just like the hen bird which has to sit long hours on an exposed nest, also without attracting attention. The drabness is a special protection. But with butterflies special protection is not always obtained by being inconspicuous. Sometimes it pays to be noticed but to look unpleasant; this particular kind of protection needs explaining.

Many butterflies (all the Danaids, for instance, and several others) have an unpleasant taste and advertise this fact by warning coloration. The Yellow Tiger butterflies and some Tiger moths are yellow and black; others are blue-black and white. By conforming to these and one or two other colour schemes, and by other superficial similarities, these poisonous species manage to be easily recognized, and avoided, by insectivorous birds.

But some quite wholesome butterflies have seized the chance to adopt false 'warning colours' and share the immunity enjoyed by the truly poisonous; and the extent of adaptation is astonishing. One female form of the Common Mormon (*Papilio polytes*) resembles the male, but the commoner female form is entirely different: this imitates the poisonous Common Rose (*Atrophaneura aristolochiae*) not only in appearance (see Plate 4, where both species are illustrated) but also in flight.

But these are at least both butterflies of the same family, both Swallowtails. On Plate 2, poisonous models on the left, mimics on the right, are shown instances more extraordinary. Fig. F, the Common Mime, is proved by every essential detail of anatomy and life-history to be a Swallowtail, and quite unrelated to the poisonous Danaid, the Blue Glassy Tiger, on its left. Fig. G is the female of a Pierid, the Wanderer, the very different male being illustrated with others of its family on Plate 7; again it is unrelated to the Yellow Glassy Tiger to its left. Both sexes of the Courtesan (a Nymphalid) are shown beside those of the poisonous Magpie Crow. Finally—strangest example of all—we show both sexes of the moth, *Cyclosia pieridoides*, alongside the Smaller Wood Nymph, which the female moth mimics.

As her greater need leads one to expect, it is usually the female which shows the greatest degree of adaptation. In the genus *Chilasa* (to which the Common Mime belongs) both sexes are mimics, but this is the exception. And there is clearly good reason that Batesian mimicry (as mimicry of poisonous by wholesome butterflies is called, after the naturalist H. W. Bates) should tend to be used only by the female. Bluff which is too frequent defeats its own end; if insectivorous birds learn that a *large percentage* of butterflies with warning coloration are good to eat, the warning ceases to be effective. Thus the species gains by having males unlike the females. A very common species gains, moreover, by having females not of a single form mimicking one species of poisonous model, and thus weakening the warning, but females of several forms mimicking several different poisonous species. When a species has more than two distinct forms ('dimorphism') the phenomenon is known as polymorphism.

PLATE 1. A. Rajah Brooke's Birdwing ♂; B. *Troides amphrysus ruficollis* ♀.

PLATE 1

PLATE 2

Classification. Together, butterflies and moths form the Order Lepidoptera, 'scaly-winged' insects. The 'scales', the fine particles of coloured dust which we find on our fingers when we handle the wings too roughly, distinguish butterflies and moths easily from other insects. In Fig. 5 are shown the structure of a Danaid butterfly's wings, the numbering of the veins and spaces, and the terms used for different parts of the wings. Alongside is a diagram of the butterfly's body. Like most other insects, a butterfly has its body divided into

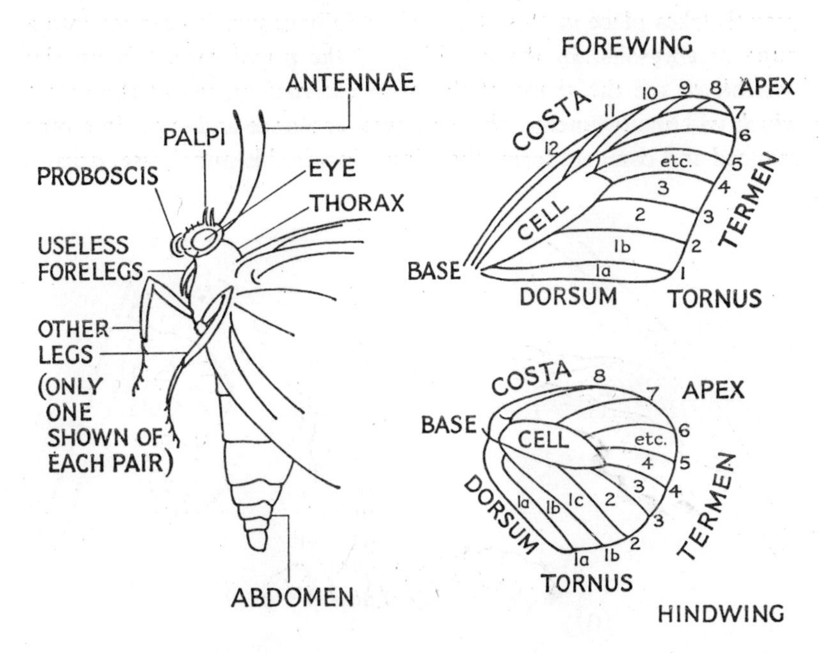

Fig. 5. Diagrams of the anatomy of a butterfly's body and wings, to explain the descriptive terms used in the text.

three fairly distinct parts: foremost, the head, with a pair of feelers or antennae, two compound eyes, and various mouthparts, chief of which is the 'proboscis', a tube through which the butterfly sucks its food and which is coiled up, when not in use, between two palpi; next the thorax containing muscles to work the limbs—four wings and six legs—which are attached to it; and last, the abdomen, which contains most of the breathing and digestive organs, and also the genital organs.

PLATE 2. A. Blue Glassy Tiger ♀; B. Yellow Glassy Tiger ♀; C. Magpie Crow ♂, D. ♀; E. Smaller Wood Nymph ♀; F. Common Mime; G. Wanderer ♀; H. Courtesan ♂, I. ♀; J. *Cyclosia pieridoides* ♂, K. ♀.

Most butterflies feed on the nectar of flowers, the excretions of aphids (honeydew), and the juices of fruit and decaying vegetable matter. They thus sustain themselves from day to day and replenish their energy. But they do not grow: all the feeding for growth has been done earlier.

Butterflies are among those insects which undergo a 'complete metamorphosis'. On emergence from the eggs, that is to say, they are grubs or larvae quite unlike the winged insects they will one day become. The larva feeds intensively, and the whole of the insect's growth takes place in this stage; when fully grown it changes into a pupa or chrysalis. In the moulding of the pupal case it is usually possible to see the shape of the main external organs of the insect which will finally emerge: the legs, eyes, antennae and even, in a very reduced and compact form, the wings. Inside this pupal case, usually

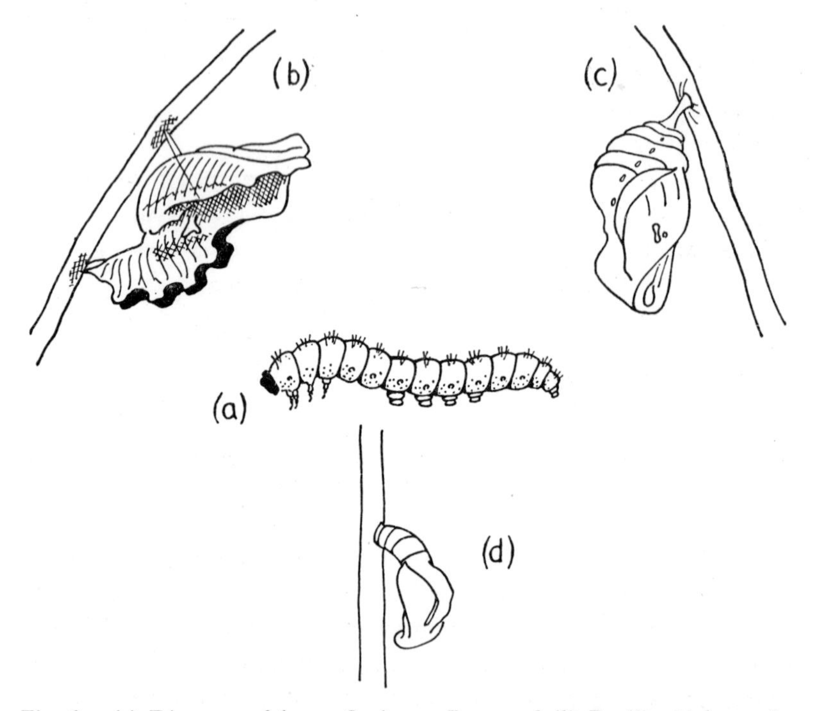

Fig. 6. (*a*) Diagram of butterfly larva. Pupae of (*b*) Papilionid butterfly (*Atrophaneura*), (*c*) Danaid butterfly (*Euploea*), and (*d*) of a Lycaenid (*Pratapa*) butterfly.

motionless but for an occasional wriggle, the insect in its final form develops; at last it splits the case and emerges.

A typical butterfly larva has a head with compound eyes, unrecognizably undeveloped antennae and a mouth quite unlike that of the butterfly itself, a mouth consisting of sharp mandibles which work sideways, biting off and chewing up morsels of leaf. Three 'thoracic segments' follow, each bearing a pair of true legs which correspond to those of the adult insect. Then follow the abdominal segments, the third, fourth, fifth, sixth and tenth of which each bears a pair of fleshy 'claspers'. Some moth larvae have only three, some only two, pairs of these claspers.

Although butterflies and moths are easily distinguished from other insects, they are not always easy to distinguish from each other. There are, indeed, more fundamental differences between some groups of moths than between moths and butterflies. The best guide is the shape of the antennae. If these are fairly slender through most of their length, but also fairly rigid, and end finally in a club, the lepidopteron is certainly a butterfly. All moths have antennae which, however shaped otherwise (they may be thickened near, but not quite at, the end), taper to a point finally; most have antennae which taper more or less flexibly throughout. But if no moth's antennae are club-ended, some butterflies have antennae which, after thickening, taper to a curved point. These all belong to the Skipper family, the Hesperiidae, and in practice they are easy to recognize on general appearance (Plate 20).

Most butterflies rest with wings erect above their backs, showing the underside, which is often intricately marked; most moths rest with wings flattened or folded back, displaying the upperside of the wings, sometimes the upperside of only the forewings; the undersides of moths, as we might expect from the fact that they are hidden, are often virtually unmarked. Most moths fly by night, most butterflies by day. But to these rules there are many exceptions, particularly to the last: rarely do butterflies remain active after dark, though many fly in the dusk, but the moths which fly by day, indeed in bright sunlight, are numerous. After a little experience the collector will not often be in doubt, and will indeed usually succeed in distinguishing butterfly from moth long before he has netted it; when doubts do occur, the shape of the antennae will prove the best guide.

On the basis of differences in structure, which in the main correspond to essential differences in the early stages, Malayan butterflies are divided into ten families. One, containing a single uncommon species, is omitted from this book.

The collector will quickly learn to assign most of his captures correctly, and to disregard some of the deceptive appearances we discussed under the heading of *Mimicry*. The beginner who is impatient that a butterfly so similar to a Danaid as the Common Mime (*Chilasa clytia*) should be classed as a Swallowtail, will at least see the absurdity of putting females of the Wanderer (*Valeria valeria*) with the Danaids (because they so cleverly mimic the Yellow Glassy Tiger) and males of the *same species* with the Pierids, where they obviously belong.

With this warning, that basic anatomical structure is the real criterion, we have introduced each family with a brief description of the general, more superficial, characteristics.

The order of families is not that which is preferred by the systematic biologist, starting with the more primitive butterflies and proceeding to the more highly evolved, but the one customarily followed by collectors.

DESCRIPTIONS OF MALAYAN BUTTERFLIES

PAPILIONIDAE The Swallowtails and Birdwings

Not all species possess 'tails', but the fact that many do accounts for the usual English name; the long and pointed forewing increases the likeness to a swallow, and this length (forewing termen longer than dorsum) is a general characteristic, and sometimes, as in *T. brookiana* and *A. aristolochiae*, very obvious.

Most Swallowtails are large insects, some magnificently coloured, usually one or two bright colours on black (Plates 1, 3, 4, 5, 6A, B and D, and 2F). Some seem to use the forewings for propulsion and the hindwings for steadiness and steering; most, however, have a strong, irregular, swinging flight. The males of many species are found on seepages on roads and river-banks.

Eggs spherical or dome-shaped; larvae with a distinctive fleshy forked 'horn' hidden behind the head, which is extruded when the insect is alarmed and which produces an unpleasant smell. Some young larvae resemble bird-droppings; most are fairly smooth, at least in later stages; but those of one group (*Troides* and *Atrophaneura*) have short, thick, fleshy spines. The head is small, and many have the body at about the third and fourth segments thickened. The pupa is always attached at the base to a cremaster or pad, and held more or less upright by a silken girdle. The adults use all six legs for walking.

RAJAH BROOKE'S BIRDWING (*Trogonoptera brookiana albescens*)
Plate 1A ♂[1]

Wallace first discovered this species, the most striking and handsome of all our butterflies, in Borneo; hence its name. Two races are found in Malaya; in one, *T. brookiana albescens*, the female is much paler than the male, the green plume markings on the forewings are less sharply defined and, towards the apex, they merge into a whitish patch. Similarly on the hindwing, the green colouring around base and cell is less distinct, and there is a series of pale submarginal spots, broken by the veins—spots which are absent in the male. The race is found in primary jungle at various places in Pahang, Perak and Selangor; the males are seen congregated at seepages or on river banks. The female is usually found higher on the hills, and is regarded as much rarer than the male, some observers having put the ratio of females to males as one in a thousand! Similar conclusions as to the scarcity of the females of other species, which collectors have drawn from observations particularly at and around river banks and seepages, which are known to attract males almost exclusively, have often been modified when the species has been observed in its actual breeding habitats, and when the insect has been bred. As *T. brookiana* has not been bred, it is idle to speculate on its true sex ratio.

In the swamp jungle of East Johore is found the other race of Brooke's Birdwing (*T. brookiana trogon*). In this subspecies (also found in Sumatra), the female is less pale than *albescens*, and thus is more like the male. In Johore, butterflies have been taken not at seepages but at flowers, and as many females have been observed as males.

[1] The signs ♂ and ♀ denote male and female respectively.

With the growing interest in natural history amongst Malayans, the earlier stages of so conspicuous and famous a butterfly will almost certainly be discovered soon, perhaps before this book is published.

THE COMMON BIRDWING (*Troides helena cerberus*) Plate 3A ♂

This insect is much more widely distributed than the preceding, and although essentially a jungle butterfly, it is sometimes seen over gardens and cultivated country, often at a considerable height from the ground. It is found occasionally even in Singapore, although the food-plant, *Aristolochia tagala*, is not found there, except in the Botanic Gardens where, however, the Common Birdwing has been known to breed.

The males, like the females, which are just as numerous, are usually captured visiting flowers.

Eggs are laid singly on *A. tagala*, and doubtless other species of *Aristolochia*, a few feet from the ground. The larva is grey with a white saddle mark, and has fleshy spines. It feeds for about four weeks, and then pupates. About three weeks later the insect emerges.

The females can be distinguished by a series of large black sub-apical spots on the hindwing. Exceptionally a few such spots are to be found on the males, and the specimen illustrated shows an unusually large number. But the completeness of these markings gives the female a quite different appearance, and once seen it is easily recognized.

Several other species are somewhat similar to *T. helena*. Of these *T. amphrysus ruficollis* (Plate 1B ♀) is most likely to be met with. The female can be distinguished from that of *T. helena* by the fact that the submarginal spots on the hindwing are separate in *T. helena*, but joined in *T. amphrysus*. Both sexes of *amphrysus* can be distinguished from *helena* by the paleness of the distal end of the cell and of the vein stripes (yellow in the male, white in the female) between the end of the cell and the apex of the forewing. *T. amphrysus* also feeds on *Aristolochia* spp.

THE COMMON ROSE (*Atrophaneura aristolochiae asteris*)

Plate 4A ♀, 4D underside

This beautiful insect, and other *Aristolochia*-feeding Swallowtails of the same group, are distasteful to most insectivores, and exhibit warning coloration. It pursues its characteristic and steady carefree flight as if interference from an enemy were unthinkable (thus falling an easy prey to that unexpected predator, the butterfly collector), and it provides a model for at least one innocuous mimic: the female-form *romulus* of *Papilio polytes* (see page 20).

The larva has thick fleshy tubercles, and is greyish red with a white saddle mark: much like the larvae of the *Aristolochia*-feeding *Troides*, but smaller and redder. It feeds on several species of wild *Aristolochia*, including *A. tagala*.

It is not uncommon in the lowland forests throughout Malaya. In Singapore it is unlikely to be met with, except in the Botanic Gardens.

THE COMMON MIME (*Chilasa clytia clytia*) Plate 2F (form *dissimilis*)

This insect is one of a group, each species of which mimics one or more poisonous Danaid species. Although widely distributed in S.E. Asia, it is not found throughout Malaya; in a few places, however, it is abundant. In Singapore it is probably the commonest 'Swallowtail'. There only the striped form, f. *dissimilis*, is found—a mimic of *Danaus vulgaris*. If, as it has been suggested, the species is a new introduction, it may follow the lead of several other common garden and roadside butterflies and spread northwards through the peninsula. In Kedah and the Langkawi Islands a dark form, called f. *onpape*, mimic of one of the *Euploea* species (the 'Black Crows': see pages 31ff), is found as well as the striped form.

In Singapore the larva feeds on cinnamon, always resting conspicuously on the upper side of the leaves. Like a bird-dropping when young, it later changes to white, yellow and black; finally the full-grown larva, very handsome and conspicuous, is black, creamy white and red. The pupa exactly resembles a broken-off twig, the posterior abdominal segments being shaped round the branch to which the pupa is attached, so that it simulates a natural growth.

The larva is so striking, and can so often be found at a convenient height on garden hedges, that where the insect occurs it will be one of the first to be bred by the collector. But the life-history, with its alternation of discreet camouflage and outrageous bluff, is an astonishing one, and the collector will go far before he finds another to equal it in interest.

The Lime Butterfly (*Papilio demoleus malayanus*) Plate 4F

This, a familiar butterfly of gardens and villages, is one of a group all of which have remarkably similar caterpillars which feed on kinds of citrus. *P. demoleus* is the commonest: almost every small lime bush seems to have larvae.

The eggs are usually laid upon young leaves; the larvae first resemble bird-droppings, later becoming green with grey oblique markings which effectually disguise them. The pupa, like those of many Papilionid and Pierid butterflies, is able to adapt its colour, within limits, to its surroundings.

Common and widely distributed in S.E. Asia, New Guinea and Australia, this insect is not found in Java, Sumatra or Borneo.

The Red Helen (*Papilio helenus helenus*) Plate 3B ♀

This large black and white Swallowtail may often be seen swinging swiftly but unevenly along jungle paths and roads on the hills. On the plains it is less common.

It is said to feed on *Zanthoxylum* and *Citrus* and to have a life-history similar to that of the other citrus-feeding Swallowtails.

P. helenus is not found in Singapore; but the rather similar *P. iswara* occurs in the catchment area wherever patches of primary forest remain. It is fond of *Mussaenda* and *Saraca* blooms, and may be caught as it pauses to hover over the flowers of these stream-side trees.

P. iswara is found throughout the peninsula, and may be recognized by the larger white patch on the hindwing, extending from space 4. In *helenus* it is confined to spaces 5, 6 and 7.

The Common Mormon (*Papilio polytes romulus*)

Plate 4B ♀-form *cyrus*; 4E ♀-form *romulus*

The English name is derived from the fact that in India and Malaysia there are two distinct forms of the female: f. *cyrus* is similar

PLATE 3. A. Common Birdwing ♂; B. Red Helen ♀.

PLATE 3

PLATE 4

to the male, except that it bears a red lunule in space 1a of the hindwing above, which the male lacks; f. *romulus* is a mimic of *A. aristolochiae*, but although the pink colour varies in extent and in brightness, it never extends to the body itself. It is a good mimic, nevertheless, imitating the flight as well as the appearance of the poisonous model. When *polytes* is bred in Singapore, the numbers of *cyrus* and *romulus* females seem to be about the same. But *A. aristolochiae* is scarce in Singapore and the mimicry serves little purpose; where *A. aristolochiae* is common, it is said that the mimics outnumber the *cyrus* females.

P. polytes is nearly as common as *demoleus*; and the larva, which also feeds on citrus, is almost indistinguishable, but the pupa is broader and more angled at the 'waist'.

THE GREAT MORMON (*Papilio memnon agenor*)

Plate 5A ♂, 5B ♀-form *distantianus*

The males of this large and conspicuous butterfly are seen almost everywhere, in gardens, open country and forests; females are less common, but the discrepancy is probably more apparent than real, since from batches of larvae, females are bred as frequently as males.

No less than five forms of the female are met with in Malaya. Commonest on the mainland is f. *distantianus*, which is illustrated. Of the other forms, those usually seen are f. *butlerianus* and f. *esperi*. These are both tailless like the male, but the forewings have the same prominent red spot at the base as in f. *distantianus* and are paler than the hindwings and without blue scales. In f. *esperi*, the forewings have a large pale patch near the apex, while in f. *butlerianus*, a pale patch appears near the tornus. Of the two remaining forms one has been taken both in Singapore and the Langkawi Islands, the other only in Singapore. They differ from forms *esperi* and *butlerianus* in having hindwings without blue scales, but with large white areas between the veins in spaces 1a, 2, 3, 4 and 5, which give the insects a quite different and much lighter appearance. They differ from each other just as *esperi* and *butlerianus* differ: in the placing of the whitish patch on the forewing.

The life-history is similar to that of *polytes* and *demoleus*; but the adult larvae have a more rugged shape and are, of course, much larger. They prefer large-leaved kinds of citrus, such as pomelo.

PLATE 4. A. Common Rose ♀, D. underside; B. Common Mormon ♀-form *cyrus*, E. ♀-form *romulus*; C. Banded Swallowtail ♀; F. Lime Butterfly.

THE BANDED SWALLOWTAIL (*Papilio demolion demolion*) Plate 4C ♀

This butterfly, often seen flying rapidly along jungle paths in the lowlands, is difficult to capture, and only too frequently the effort of netting it breaks one or both of the 'tails'. It may be taken at flowers, however, and is especially fond of *Saraca*, *Mussaenda* and *Ixora*.

The larva is said to feed on citrus, but the butterfly prefers forest to villages and gardens, and I have found the larvae, not on cultivated citrus, but on the jungle climber, *Luvunga scandens*. The larva bears resemblances, at various stages, to those of typical citrus-feeding Swallowtails (*P. demoleus*, *memnon* and *polytes*), but there are important differences. First, the eggs are laid not singly, but six or more together one on top of another in a rod projecting from the stem of the food-plant. Then, as one might expect, the young larvae are gregarious; in the 'bird-dropping' stage, they are paler than those of *demoleus* or *polytes*; later they are a bluer green, and they still retain gregarious instincts. A large percentage of larvae are parasitized by a wasp. The pupa is distinctive in possessing a long dorsal spike.

THE FIVE-BAR SWORDTAIL (*Graphium antiphates itamputi*)
Plate 6A ♂

Swift in flight, this species may be mistaken for a Pierid butterfly. It is not uncommon on roads and in forest clearings, and the males are sometimes seen congregated on moist spots.

The food-plant is said to be a climber of the family Annonaceae.

A somewhat similar insect with the same rapid flight but much rarer, and frequenting not roads and seepages in the lowlands but open hilltops above 3,000 feet, is *G. agetes*. The two species cannot be confused: *agetes* is less heavily marked with black on the upper surface than *antiphates* (see Plate); and *agetes*, unlike *antiphates*, is marked on the underside conspicuously with red.

THE COMMON BLUEBOTTLE (*Graphium sarpedon luctatius*) Plate 6D ♀

Although this familiar butterfly is one of a group of *Graphium* Swallowtails, all with similar colouring, it can be confused with none of the others since the blue markings are simpler and bolder (compare the illustration with that of *G. evemon*, B).

The green larva feeds on various species of Lauraceae, seeming to prefer cinnamon, and developing in the final instar a distinctive transverse yellow bar on the third thoracic segment. It is sluggish, often feeding throughout its larval life on a single spray of leaves, then pupating on the undersurface of one of them. The pupa has a prominent dorsal spine, and is ridged lengthwise in a way which forms a perfect camouflage underneath the longitudinally-veined cinnamon leaves.

In bred specimens the turquoise colour is greener and lacks the brilliancy of specimens captured on the wing.

Another common butterfly is *G. evemon*: B on Plate 6 will serve to distinguish it from *sarpedon*. With other members of the group it can be confused more easily; but on the underside of the hindwing there is a short black bar at the costa which will serve for identification if it fulfils two conditions: it must join up with another black bar at the base of the wing, and it must not be spotted with red. If it is linked with the basal streak but is red-spotted, it may be identified as *G. eurypylus*, while if it is separate, not only from the basal streak but also from all other black markings, it is *G. doson*.

The larva of *G. evemon* is not unlike that of *sarpedon* in its earlier stages, but in the final instar it develops two prominent 'eye spots' on the third thoracic segment. One food-plant is *Artobotrys*.

One other *Graphium* species must be mentioned since around villages and gardens it is almost as abundant as *sarpedon*. This is the TAILED GREEN JAY, *G. agamemnon*. Usually larger in size than other members of the *sarpedon* group, it has spots which are green (not bluish) and more numerous and separated, giving it a distinctively speckled appearance. The larva of *agamemnon* feeds on soursop (durian blanda) leaves, *Michelia champaca*, and various other plants. The half-grown larva has a curious rectangular pale dorsal patch on the anterior abdominal segments. Otherwise it is similar to other *Graphium* larvae: green in colour, and very broad at the third thoracic segment.

PIERIDAE THE WHITES AND SULPHURS

Medium-sized butterflies containing white, yellow or orange coloration, sometimes with bolder colours on the underside of the hindwing, and usually with black veining or edging to the wings, which is more extensive in the female (Plate 6, C, E and F, Plate 7, and Plate 8A and B).

Eggs usually recognizable by slenderness and height. Larvae usually some shade of green, often marked longitudinally; long and cylindrical, rather plain—no 'horns' or 'tails' but sometimes with short, sparse hairs. Pupae usually green, fastened anally, and held erect by a girdle (like those of the Papilionidae). Adults with all six legs functional.

THE PAINTED JEZEBEL (*Delias hyparete metarete*)

Plate 6E ♂ underside, F ♀

The colours of this butterfly—white with strong black veining, contrasted with the bright yellow and red on the underside—together with its weak flight, are in this instance true warnings of its unpleasant taste.

At no stage of its life is any attempt at concealment made: the eggs are laid, about a dozen at a sitting, spread over an area the size of a five-cent coin, on a *Loranthus* (mistletoe) leaf. The gregarious larvae are an unpleasant deep oily yellow, with black heads and sparse black hairs. The pupae are a polished milk-white with black spots, and are conspicuous—sometimes several together—on the upper surface of a *Loranthus* leaf.

Uncommon on the hills, *D. hyparete* is abundant on the plains. On the hills the commonest *Delias* is *D. ninus*, the MALAYAN JEZEBEL (Plate 6C), which often occurs in localized colonies.

THE ORANGE ALBATROSS (*Appias nero figulina*) Plate 7D ♂

The life-history of this beautiful and widely distributed butterfly is unknown. The males may often be taken in some numbers on sandy river-banks and seepages on the plains. The females, recognized by the much darker veining at the apices and termens, probably have a migratory tendency, since they are seldom found with the males, but are usually taken singly at flowers in the jungle or on hilltops. They are considered to be rare. In Singapore the butterfly appears to be only an occasional visitor.

In collections, the beautiful orange-vermilion darkens in time to deep crimson.

PLATE 5. A. Great Mormon ♂, B. ♀-form *distantianus*.

PLATE 5

PLATE 6

THE CHOCOLATE ALBATROSS (*Appias lyncida vasava*) Plate 7A ♂, B ♀

This insect is abundant and widely distributed in the forested plains of Malaya, except in Kedah where it is not found at all, and in Singapore where it is uncommon and seems to be a seasonal migrant. The English name describes the wing borders on the underside; apart from these broad chocolate borders, the hindwings beneath are sulphur-yellow.

Corbet gives *Crataeva religiosa* and *Capparis micracantha* as food-plants. In Singapore I have seen a female ovipositing on *Gynotroches axillaris*.

A rather similar butterfly, but distinctly smaller and lacking the chocolate border and the sulphur-yellow beneath (except for a narrow dash of yellow on the costa), is *Appias libythea*, the SMALL BLACK-VEINED ALBATROSS. Until a few years ago, apparently, this insect had made only occasional appearances in Malaya; it then colonized Singapore and, since the war, has spread rapidly northwards, and is now one of the commonest roadside butterflies, at least on the west. The abundance of the larval food-plant, *Cleome ciliata* (the wild cat's-whisker), which springs up wherever ground is cleared and grass cut back, has doubtless made the butterfly's spread possible: where roads have gone, *libythea* has been able to follow.

THE GREAT ORANGE TIP (*Hebomoia glaucippe aturia*) Plate 7C ♂

The figure will identify this, the largest Pierid found in Malaya, beyond doubt. The rarer female, as in almost every Pierid species, is more heavily marked with black. The butterfly has a strong flight and is difficult to capture; but the males may be taken at moist spots on river banks. The larva feeds upon *Crataeva religiosa*.

The butterfly is not uncommon on the plains of the peninsula mainland, and on some of the islands, but is not found in Singapore.

THE WANDERER (*Valeria valeria lutescens*) Plate 7E ♂, Plate 2G ♀

Like *Hebomoia* the genus *Valeria* contains but one Malayan species, and the males of the Wanderer, which are not uncommon at forest edges, quarries and similar localities on the plains and foothills, cannot be mistaken for any other butterfly. The female, however, is a

PLATE 6. A. Five-bar Swordtail ♂; B. *Graphium evemon*; C. Malayan Jezebel; D. Common Bluebottle ♀; E. Painted Jezebel ♂ underside, F. ♀.

remarkably accurate mimic of *Danaus aspasia*, the Yellow Glassy Tiger, as may be seen by comparing the two, Figs. B and G on Plate 2. It also imitates the rather slow weak flight of the Danaid butterfly.

The males are swift in flight, and are difficult to capture unless settled on the ground. The butterfly is an occasional visitor to Singapore, usually seen only in May or June. On Pulau Ubin, near by, it is commoner.

THE LEMON MIGRANT (*Catopsilia pomona pomona*) Plate 8A ♂, B ♀

There are several distinct forms of the Lemon Migrant, and although some are variable, others are constant, and it is possible that two species are included under one name; some years ago, indeed, they were considered distinct.

Those illustrated on Plate 8 are both of the type which has always been known as '*pomona*'. These have less black on the wings and, although this does not appear on the upperside, at least a little red pigment. On the male, this is seen in red rings round the cell-end spots on the underside of the hindwing, which is otherwise nacreous; and in the female, in the slightly orange tint of the underside which may be marked, sometimes extensively blotched, with dull red; in both sexes, reddish colour is noticeable on the antennae and, in fresh specimens, on the eyes.

The other form, until recently known as '*C. crocale*', has no red or orange pigment. It is commoner than '*pomona*'. The males are constant: the upperside is like that of '*pomona*', but the narrow black of the apex extends down the termen; the underside of the hindwing is lemon-yellow, chalky not nacreous, and with no spots. The females are very variable: seldom a flat even yellow over the whole upperside, as on the '*pomona*' female illustrated; never with red spotting or blotching on the underside; and usually with a blackening of the costa on the upperside of the forewing, narrow at the base, thickest half-way along the wing—a wedge-shaped mark which may extend to and include the cell-end spot.

The larvae feed on *Cassia siamea* and *C. fistula*, the young larvae hiding along the leaf-veins and the adult larva sometimes spinning two leaflets together and resting down the centre as if it were the midrib of this enlarged 'leaf'.

The insects are easy to breed; but to determine the status of the two 'species' or 'forms', experiments would have to be devised in some large breeding enclosure in which the insects could pair, and the females oviposit, under observation (see page 8 above).

Two related species, almost as common as *C. pomona*, are *C. scylla*, the ORANGE MIGRANT, and *C. pyranthe*, the MOTTLED MIGRANT. Neither is so variable as *pomona*, and both are easily recognized. *C. pyranthe* is very pale greenish or greyish white, with the whole of the hindwing beneath and the apex of the forewing minutely lined, giving a mottled appearance. The larvae seem to prefer *Cassia occidentalis*, a small tree or shrub of shore and waste ground. *C. scylla*, as the English name indicates, is distinctively orange—much richer in colour than the reddish forms of *pomona*, and usually smaller. Most of the orange is on the underside, but *scylla* is quite as distinctive above: the forewings above are milky white, faintly tinted with orange, and the hindwings are invariably deeper in colour. This contrast is very marked in the male, fainter but still quite distinct in the female. *C. scylla* is also a butterfly of roadsides and waste patches, the larva feeding on *Cassia tora*.

THE COMMON GRASS YELLOW (*Eurema hecabe contubernalis*) Plate 7F

The little fluttering Grass Yellows, with their bright yellow wings and neat black borders, are perhaps the most familiar of Malayan butterflies. They are abundant and, despite their small size, conspicuous.

E. hecabe is the commonest of the genus; it is also deceptively variable, especially in size. After *hecabe*, the most plentiful is probably *E. sari* which may be recognized at once by the prominent undivided brown area at the tip of the forewing beneath (see Plate 7G).

Eurema larvae feed on *Pithecellobium*, *Caesalpinia*, *Cassia*, *Cratoxylon* and doubtless many other plants.

A species which might at first sight be mistaken for a member of the genus is the TREE YELLOW, *Gandaca harina*. When captured and examined, however, it will be seen to have no markings either above or below, except for a narrow black apex, slightly broader in the female, on the upperside of the forewings. *G. harina* is hardly a garden butterfly; but it is common in primary and secondary growth. The larva (a bluer green than most Pierid larvae) feeds and sometimes pupates inside rolled leaves of *Ventilago oblongifolia*.

DANAIDAE The Milkweed Butterflies, Tigers and Crows

The Danaidae are considered to be the most highly developed of all butterflies. They are medium-sized to large; all have an unpleasant taste or smell; and all fly in a leisurely manner, confident that few insectivores will have the hardihood to attack. A few birds are too young to know, too hungry to care, or have digestive systems as admirable as those of the Bee-Eaters; but most leave well alone. Danaids are, moreover, remarkably tough: many a Danaid butterfly, firmly pinched or otherwise killed, and papered by the entomologist, has later, when the paper has been unfolded for inspection, scrambled to its four feet, spread its wings and disappeared through the open window. The males have secondary sexual characters in the form of brands and patches of scent scales on the wings. They also have a pair of feathery brushes which they are able to extrude from the abdomen, and which are normally used in the process of courtship to brush the scent from a patch of scent scales on the hindwing; the largest of the black spots on the hindwing of *Danaus chrysippus* (Plate 8D) is in fact this patch of scent scales. In many species, however, the brush is extruded when the insect is alarmed. This disconcerting trick and the unpleasant smell and taste are protections not only for the species but also, since the insects are capable of reviving after what would to another butterfly be a fatal injury, for the individual.

The family is remarkably conservative in colouring and type of marking, being divided into three or four groups, each containing a number of very similar butterflies: yellow-brown, white and black; plain white and black; blue or greenish blue and black; and so on (Plate 2A, B, C, D and E, Plates 8 and 9, and A of Plate 10). The value of this conservative colouring (Müllerian mimicry) and the fact that these butterflies serve as models for many unrelated species is discussed on pages 11 and 12 above.

Such Danaid eggs as I have seen are oval, attached at one end, and

PLATE 7. A. Chocolate Albatross ♂, B. ♀; C. Great Orange Tip ♂; D. Orange Albatross ♂; E. Wanderer ♂ (♀ see Plate 2 G); F. Common Grass Yellow; G. *Eurema sari sodalis* underside.

PLATE 7

PLATE 8

appear smooth and yellowish. The larvae are conspicuous and distinctive, with three or four pairs of long black tapering spines. In colour some *Euploea* larvae are oily yellow and black; some *Danaus* larvae are brightly coloured, with transverse stripes of black, blue and yellow. They feed on latex-bearing plants, such as *Ficus* and *Asclepias*. The pupae are attached to a silken pad by minute hooks on the anal segment, and hang head downwards without other support; they are rounded and compact (see text figure of *Euploea* pupa page 14), with opalescent or sometimes brilliant metallic colours. The adults, like those of other families which have anally suspended pupae, have the forelegs undeveloped, and useless for walking.[1]

THE PLAIN TIGER (*Danaus chrysippus alcippoides*)

Plate 8D ♂ (form *chrysippus*)

Though not abundant, this butterfly is familiar enough in cultivated areas: the conspicuous and beautiful larva feeds on *Calotropis*, which is sometimes grown for medicinal purposes by Indian communities, and on *Asclepias*. Although, therefore, the collector may catch only occasional glimpses of single specimens disappearing over hedges in cultivated and suburban areas where a determined chase is embarrassing or impossible, he will find the larvae on the underside of *Calotropis* leaves, and breed as many as he wants. The larva is marked with black, pale blue and yellow, and has six of the long black spines typical of Danaid caterpillars. The pupa is almost as beautiful; it has the colour and lights of a very pale opal, with a few black spots.

The insect illustrated is form *chrysippus*, and is typical of most insects found in Penang, Kedah and Singapore. Through the rest of Malaya, form *alcippoides*, which has the hindwing largely white, the yellow-brown colour remaining only narrowly along the black margin, has replaced the plain *chrysippus* form.

THE BLACK-VEINED TIGER (*Danaus melanippus hegesippus*) Plate 8C ♂

This butterfly is very common everywhere on the plains, and occasionally ascends the hills. The hindwing, entirely white ground with black veins, will serve to distinguish it from any other butterfly of the same group. The slightly larger *D. genutia* is also tawny-coloured and

[1] The remarkable fact that Danaids, and only Danaids, are attracted to certain dried plants, is commented upon on page 7 above.

PLATE 8. A. Lemon Migrant ♂, B. ♀; C. Black-veined Tiger ♂; D. Plain Tiger ♂ (form *chrysippus*); E. Common Tree Nymph ♂.

veined with black, but the black veining is less conspicuous, and the hindwing is coloured like that of *D. chrysippus*: a pale hindwinged form is replacing the tawny hindwinged form over most of Malaya, but the hindwings show at least a trace of tawny near the border. As any collector taking *genutia* will certainly have specimens of the much commoner *melanippus* for contrast, confusion is unlikely.

THE YELLOW GLASSY TIGER (*Danaus aspasia aspasia*) Plate 2B ♀

The illustration will serve to distinguish this butterfly from any other Danaid, but it is so closely mimicked by the female of the Wanderer, *Valeria valeria lutescens* (Plate 2G), that the two are easily confused when seen on the wing. *D. aspasia* is much commoner, however; it is a familiar and fairly widely distributed forest butterfly on the mainland, but rare in Singapore.

THE LARGE CHOCOLATE TIGER (*Danaus sita ethologa*) Plate 10A ♂

This insect occurs only on the hills, but above 4,000 feet it is sometimes common and, because of its size and colour, conspicuous. The blackening on the reddish-brown border of the hindwing indicates the scent scales of the male. The larger female is without this blackening.

A smaller insect, also frequenting the hills, but coming much lower, is the COMMON CHOCOLATE TIGER, *D. melaneus*. The hindwing border is dark chocolate, and the general appearance of the insect is blackish rather than reddish brown.

THE BLUE GLASSY TIGER (*Danaus vulgaris macrina*) Plate 2A ♀

This insect is common everywhere in Malaya. It is rather similar to an almost equally common insect. *D. agleoides*, but the two may easily be distinguished by the forewing cell: *agleoides* having only a thin longitudinal black line on the pale cell-streak, and *vulgaris*, as the illustration shows, having the pale cell-streak crossed obliquely by a curved black bar.

The larvae of both are said to feed on *Gymnema*; but in Malaya they probably feed on other Asclepiadaceous plants as well.

THE SMALLER WOOD NYMPH (*Ideopsis gaura perakana*) Plate 2E ♀

This butterfly is common on the forested hills, and much less common on the plains. I have not heard of its being taken in Singapore; nor have I seen it in South Johore; otherwise it appears to be widely distributed. The male is darker in colour than the female, and has narrower wings.

THE COMMON TREE NYMPH (*Idea jasonia logani*) Plate 8E ♂

Several species of *Idea* are found in Malaya, but *jasonia* seems to be the commonest and, with its elongated wings, it is not the least beautiful. *I. lynceus* is not quite so common, and it is so similar that no infallible means of distinguishing it may be given here. Two rarer species may, however, be distinguished easily: *I. hypermnestra* has much more rounded and much whiter wings than *I. jasonia* (which is always dusted, sometimes quite darkly, with grey); while the very rare *I. leuconoë* has yellow-tinted wing bases. The last has a special interest: it is one of the few butterflies which seem to be confined to mangrove swamps.

The length of wing and proportion of wing area to size of body give species of *Idea* a strangely alternating flight of fluttering and gliding, which is almost a parody of the flight of Danaids in general. Their timeless and ghostly movement high up in the shade of the canopy is one of the memorable sights of the Malayan jungle, and to my mind the Malay nickname of 'surat' is strangely inadequate.

THE STRIPED BLUE CROW (*Euploea mulciber mulciber*) Plate 9A ♂, B ♀

This beautiful butterfly is fairly common in Singapore; elsewhere in Malaya it is abundant. As in other *Euploea* species the dorsal margin of the forewing of the male is extended in a curve, giving the wings a more rounded appearance than those of the female; they have long feathery brushes which they extrude when they are caught; and also an extended area of specialized scent scales on the upperside of the hindwing (instead of the much more restricted pad or vein of such scales in *Danaus* males). Unlike most other *Euploea*, however, *E. mulciber* has no sex-brand in space 1b of the forewing.

The striping of the female is reminiscent of a *Danaus*, and most of the *Euploea* females have rather more distinct white striping on the hindwing than the males. *Nerium* (oleander) and *Aristolochia* are given as food-plants; I have found the larva on two species of *Ficus*.

THE STRIPED BLACK CROW (*Euploea eyndhovii gardineri*) Plate 9D ♂

This is another common and widespread 'Crow', but not so abundant as *mulciber*. The male has a thin brand on space 1b of the forewing above; the female is a duller brownish black, with more distinct striping.

This species is the model for a very rare form of the Papilionid BLUE MIME, *Chilasa paradoxa* (the more usual form of which mimics *E. mulciber*), and for several other interesting mimics, including a moth—a relation of the *Cyclosia* moth illustrated on Plate 2 J and K. It is well worth while taking a second glance at any *Euploea* which does not appear to be quite typical.

THE MAGPIE CROW (*Euploea diocletianus diocletianus*) Plate 2C ♂, D ♀

This species also serves as a model for several rare and quite unrelated species, one of which, the female of the Nymphalid butterfly, *Idrusia nyctelius*, is shown beside it on the plate. The males are common everywhere in Malaya, on the hills as well as the plains. The females are rather rare, frequenting only the forests; but, like the males, they can be attracted by dried plants of *Heliotropium indicum*.

THE BLUE-BRANDED KING CROW (*Euploea leucostictos leucogonis*) Plate 9C ♂

This handsome insect, though usually much larger than the last species, is very variable in size. It is occasionally abundant, particularly on some hill stations, but is uncertain in appearance. The female is remarkably similar to the male, having even a blue spot to replace the male brand, but the dorsal margin on the forewing is straight.

THE DWARF CROW, *E. tulliolus*, is remarkably similar in outline and marking, but much smaller, and it lacks the forewing brand or spot in space 1b.

Of the other *Euploea* species, those most likely to be seen are the MALAYAN CROW, *E. redtenbacheri malayica*, and the SPOTTED BLACK CROW, *E. crameri*. The former is a large insect, widely distributed, and not uncommon. It is black, with spots over the distal two-thirds of both wings; the general appearance is of a fairly *even distribution* of spots and is quite distinctive. *E. crameri* is not so widely distributed

PLATE 9. A. Striped Blue Crow ♂, B. ♀; C. Blue-branded King Crow ♂; D. Striped Black Crow ♂.

PLATE 9

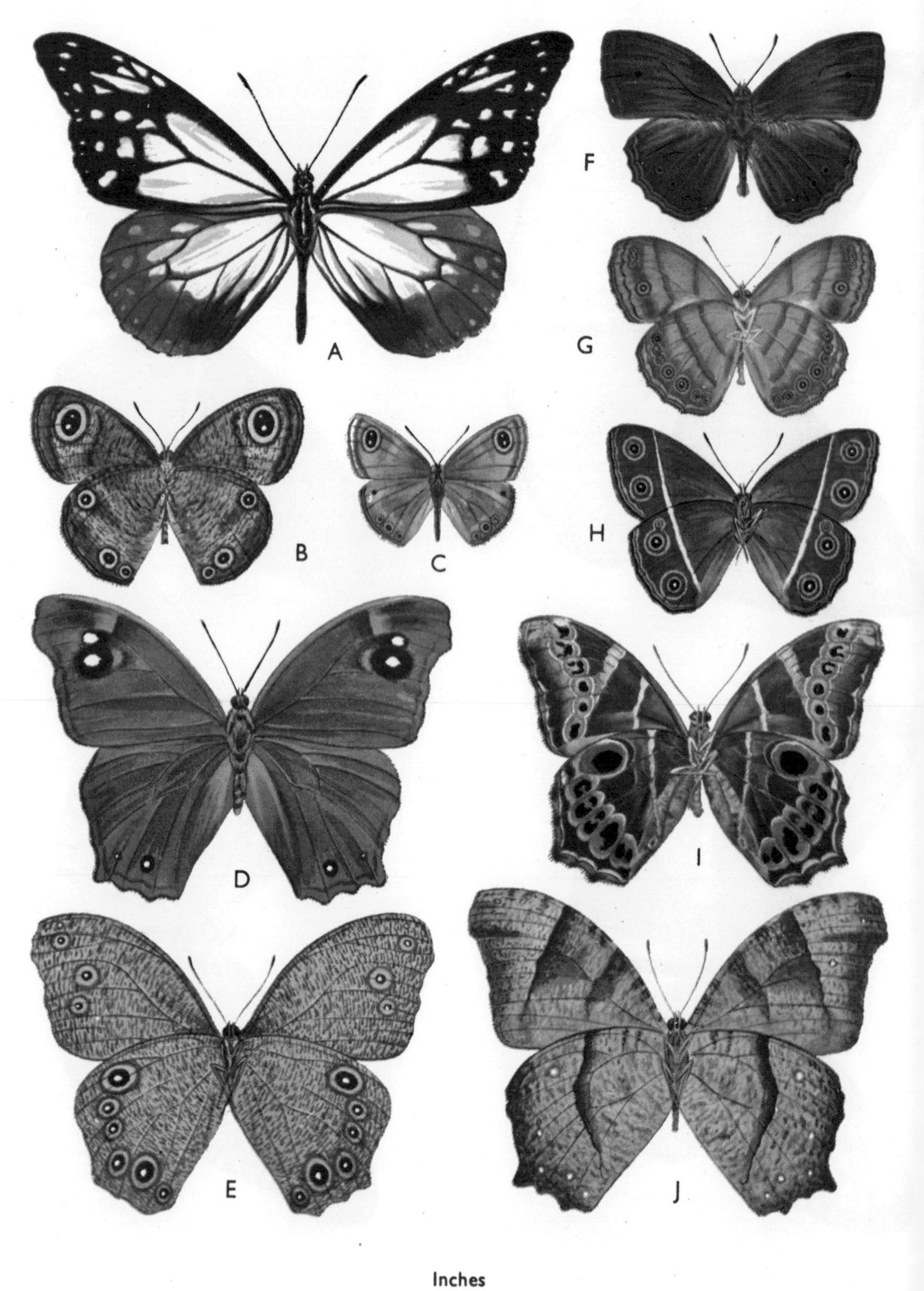

PLATE 10

an insect as *redtenbacheri*, but where it occurs, it is commoner: it is not uncommon on the Langkawi Islands and on the east coast; and it is the commonest *Euploea* in Singapore. Like *redtenbacheri*, it is black or dark brown, spotted with white; but the four largest spots are grouped, almost touching each other at the apex of the forewing, so that this species too is quite distinctive in appearance.

SATYRIDAE The Browns and Arguses

Mostly rather small butterflies; broad- and somewhat round-winged, and slow, flying close to the ground and preferring shade. Usually dull brown in colour, with numerous ocelli on the underside (Plate 10, except for fig. A, and Plate 11C, D and G). The males of many species have scent brands and brushes, and a distinct thickening of some veins at the base of the forewing.

The food-plants are monocotyledons: grasses, palms and bamboos. The early stages of some quite common species are unknown; partly, perhaps, because some Satyrids do not place their eggs deliberately on a carefully chosen food-plant, but merely drop them at random while in flight over grasses, and are thus difficult to observe. The larvae taper towards the head, which is sometimes 'horned'; and also towards the rear, which is doubly 'tailed'. Most of the pupae are anally suspended, without other support; some are formed on the ground amongst grass roots. The forelegs of the adults are not used for walking.

The Common Three-Ring (*Ypthima pandocus corticaria*)

Plate 10B underside

A poor Cinderella of butterflies, so common that no one has troubled to observe its life-history. Presumably the larva feeds on various grasses.

Slightly smaller is the Common Five-ring, *Ypthima baldus*; and smaller still is *Y. ceylonica huebneri*, the Common Four-ring (Plate 10C). All three species are abundant.

The English names, Three-ring, Four-ring, etc., refer to the number of eye-spots or ocelli on the *underside of the hindwing*, not to the total number of spots; and the two spots at the hindwing tornus, contained in a single yellow outer ring, count as only one.

PLATE 10. A. Large Chocolate Tiger ♂; B. Common Three-Ring underside; C. Common Four-Ring; D. Common Evening Brown upperside, E. and J. undersides; F. Malayan Bush Brown, G. underside; H. Nigger underside; I. Bamboo Tree Brown underside.

The Bamboo Tree Brown (*Lethe europa malaya*)
Plate 10I ♂ underside

The butterfly will readily be identified from the plate. The upperside of the male is dark brown, with a few indistinct marks on the wing borders and some blurred subapical spots. The female has a broad white band across the forewing apex.

As the name suggests, the larva feeds on bamboo, and the butterfly is not uncommon around villages. But it is seldom seen, as, unless it is disturbed, it flies only in the early morning and at the approach of dusk.

The Malayan Bush Brown (*Mycalesis fuscum fuscum*)
Plate 10F ♂, G underside

There are many species of *Mycalesis*, or Bush Browns, in Malaya; all the males have hair tufts on the costa of the hindwing above, and scent scales, against which these tufts brush, on the underside of the forewings. *M. fuscum* is not the commonest, indeed it is a forest butterfly, whereas several species of *Mycalesis* are found in gardens and on the forest edges; but it is not uncommon, and it can at once be distinguished from its greyer, drabber congeners (of which the commonest is *M. mineus*), by its ochreous underside, crossed by reddish-brown lines.

The Nigger (*Orsotriaena medus cinerea*) Plate 10H underside

This butterfly, which is very common in shady places all over the plains of Malaya, may at once be recognized from the plate. The larva is known to feed on grasses.

The Common Evening Brown (*Melanitis leda leda*)
Plate 10D upperside, E and J undersides

Why Linnaeus named this drab butterfly 'Leda' is puzzling. The least uninteresting thing about it is the variability of the underside. In lands where there are marked wet and dry seasons (the insect is found in the Old World tropics from Africa to Australia) the variation is seasonal, but in Malaya the different forms can usually be found at any season of the year.

The larva feeds on the leaves or blades of rice, and the butterfly is common around paddy fields, flying at dusk. But there must be other food-plants, as it is also common in gardens and forests scores of miles from the rice fields. In other countries it is said to feed on various grasses, sugar cane and bamboo.

THE COMMON PALMFLY (*Elymnias hypermnestra beatrice*)
Plate 11C ♂, subsp. *beatrice* (Malay Peninsula mainland)
D ♀, subsp. *agina* (Singapore)
G ♀, subsp. *tinctoria* (Langkawi Islands)

All members of the genus *Elymnias* are interesting for some degree of mimicry. In one very rare species (*E. künstleri*), illustrated in Corbet and Pendlebury's book, the male is an excellent mimic of one of the Black Crows, and the female differs so far as to mimic a species of *Idea*.

The only common species is *E. hypermnestra*, the larva of which feeds on coco-nut and other palms. It is abundant throughout Malaya, but is found in several distinct geographical races. In Kedah and the Langkawi Islands, the female (Plate 11G) has the general colouring of *Danaus genutia*, and like that butterfly, flies in the sunshine. In central Malaya, the female is similar to the male, and prefers the shade; it is not unlike *Euploea tulliolus*, the Dwarf Crow, but can hardly be regarded as a mimic since it is much commoner.[1] In Singapore and South Johore occurs a somewhat similar race, but the female is larger and has paler spots.

The TAWNY PALMFLY (*E. panthera*) is the only other member of the genus certain to be seen by the average collector. Its sharply angled hindwing and mottled underside will reveal its relationship to *E. hypermnestra*; in flight, however, and at a distance, it could be mistaken for a female, or a faded male, of *Euploea eyndhovii*.

AMATHUSIIDAE THE AMATHUSIIDS

These are large butterflies, some very large, usually dark in colour, with prominent underside ocelli and conspicuous scent brands and brushes in the males. Except that many bear brilliant metallic blue patches, or blue 'shot' wings, Amathusiids are much like gigantic Satyrids (Plate 11, A, B and E, and Plate 12).

[1] For the mimicry to be effective, the poisonous model must be commoner than the harmless mimic (see page 12).

The larvae, also like Satyrids, are centrally thickened and have two 'tails'. They feed on monocotyledons: various palms, wild banana, etc. Some are gregarious and hairy.

The insects frequent palm and rotan thickets in the darker parts of the forest, and easily elude the entomologist whose net, clothes and skin are caught up in thorns and rotan hooks if he attempts to follow. Most Amathusiids, however, can be enticed by rotting fruit.

The Common Faun (*Faunis canens arcesilas*) Plate 11A

Common in lowland forests and up to about 2,000 feet, this butterfly is usually seen close to the ground, flying in shady undergrowth. The larva is said to feed on wild species of banana.

The Yellow-barred (*Xanthotaenia busiris busiris*) Plate 11B

This easily identifiable insect has similar habits to the preceding, and sometimes frequents the same patches of forest. The life-history is unknown.

The Palm King (*Amathusia phidippus chersias*)

Plate 12A ♂, B ♀, C underside

We have all seen the occasional specimen of this butterfly settled on wall or ceiling. It is common around towns and villages, where the larva feeds on coco-nut and other palms, and although it flies at dusk or in the early hours of darkness it seems attracted to the lights or shelter of houses. It is very fond of rotting fruit; it may be, therefore, our garbage bins which are the first attraction.

The larvae feed on the underside of palm leaves, and are gregarious at least when young, when they rest in a radiating formation appearing like a circular discoloration of the leaf. They are hairy, and a slightly pinkish brown. The pupa is green, anally suspended; although large, it is gracefully shaped.

Of others in the same genus, all are rare jungle butterflies except *A. gunneryi*, the Faded Palm King. This has an underside like that of a faded and indistinct *phidippus*, and is abundant around Kuala Lumpur, but nowhere else. Corbet describes the larva as green, and this supports his contention that this species is distinct.

PLATE II. A. Common Faun; B. Yellow-barred; C. Common Pal mfly ♂, D. subsp. *agina* ♀, G. subsp. *tinctoria* ♀; E. Saturn ♂, F. ♀.

PLATE 11

PLATE 12

THE SATURN (*Zeuxidia amethystus amethystus*) Plate 11E ♂, F ♀

This insect may sometimes be seen along jungle paths at dusk, the male more commonly than the female. Both sexes come freely to rotting pineapple hung in suitable places.

Z. doubledayi is less common than *amethystus*, but may also be taken at fruit bait in the jungle. The blue patch on the hindwing of the male extends broadly to the apex, and the female has pale mauve spots instead of yellow. On the underside, both sexes are purple-washed, so that they can usually be distinguished from *amethystus* even when at rest.

THE DARK BLUE JUNGLE GLORY (*Thaumantis klugius lucipor*)
Plate 12D ♂

This is an uncommon butterfly, but one of the most beautiful. It is the only Amathusiid which bears any comparison with the brilliant Morpho butterflies of Central America, to which indeed the Amathusiids are distantly related.

Occasionally, as we make our way through the dark forest undergrowth, this lustrous deep blue butterfly will spring to life beneath our feet and, after a short flight during which the blue flashes darkly and mysteriously will disappear again completely. Its cryptic underside pattern renders it virtually invisible until once again it takes wing.

The much commoner DARK JUNGLE GLORY, *T. noureddin*, is less handsome: the male is dark purplish brown above, with the wing bases shot with a blue which is clearly visible only when the insect is in flight, not set in a collection. The female is bluer.

Thaumantis species do not come to fruit bait.

NYMPHALIDAE THE NYMPHALIDS

These butterflies show great variation in size, shape and colouring (Plates 13 to 16 and Plate 17A), and it is difficult to give any superficial character by which they will at once be recognized. Like the Danaids, the Satyrids and Amathusiids, they have forelegs which are useless for walking, and their pupae hang supported only by anal

PLATE 12. A. Palm King ♂, B. ♀, C. underside; D. Dark Blue Jungle Glory ♂.

hooks; unlike those three families, the males have no special scent brands; and unlike the last two, most Nymphalids are sun-loving butterflies.

The numerous species of Nymphalids are varied also in their habits: a few like dung and rotting fruit and prefer the jungle shade; many are typical butterflies of our gardens and open places, visiting flowers; others are butterflies of jungle clearings and edges and tree-tops. Nearly all have a strong, well-controlled flight.

The larvae too are varied, and often show a high degree of protective adaptation, either in form or in habits. Some are elaborately spined; and the food-plants, unlike those of the two preceding families, are usually dicotyledons. The hanging pupae also show an amazing diversity of form, imitating shrivelled or torn leaves, berries or empty pupal cases. Many have metallic facets or spots.

The Rustic (*Cupha erymanthis lotis*) Plate 13A

A familiar little butterfly of the forest edges, with a restless, rather weak flight.

The eggs are laid on young leaves of *Flacourtia rukam*. The larvae are green or dark brown, with small branched spines; when disturbed, they drop from the plant on a thread. The pupa is bright green with red spots.

The Vagrant (*Vagrans egista macromalayana*) Plate 13B

Not unlike *C. erymanthis* in colour, the Vagrant has sharper wing contours and an altogether more energetic flight. It is an insect of clearings, quarries and forest edges, but prefers the hills to the plains.

The larva is similar to that of the Rustic, and feeds also on *Flacourtia rukam*. The pupa is green with red and gold spots.

The Cruiser (*Vindula arsinoe erotella*) Plate 13E ♂, F ♀

This beautiful insect is common in Singapore, males more so than females, and fairly common on the plains throughout Malaya, frequenting forest roads and edges. It has a strong and lofty flight; and although many forest butterflies choose low and sheltered branches on which to lay eggs, the female of *arsinoe* may often be seen reconnoitring and ovipositing at a considerable height from the ground.

The larva is grey, with a prominent black horned mask; and from the first abdominal segment it tapers slightly towards the rear. It is rather sparsely spined. The food-plant is *Adenia*. The beautiful pupa has lengthy and curiously shaped appendages giving it the appearance of a leaf skeleton.

The almost indistinguishable *V. erota* occurs on the hills.

THE BANDED YEOMAN (*Cirrochroa orissa orissa*) Plate 13C

Most Malayan species of *Cirrochroa* resemble *Vindula* males in general appearance, but have a silvery band on the underside of the hindwing which is sometimes continued on to the forewing, and no hindwing 'tails'. *C. orissa*, with its broad black apex and yellow band across the forewing, unmatched by any such band on the hindwing, is quite distinctive. It is also the commonest, frequenting forest edges and clearings.

THE ROYAL ASSYRIAN (*Terinos terpander robertsia*) Plate 13D

This beautiful butterfly is widely distributed and is not uncommon on forest roads, around quarries, sometimes at moist spots, on the Malayan mainland. I know of no Singapore locality.

THE MALAY LACEWING (*Cethosia hypsea hypsina*)
Plate 14A ♂, D underside

Fresh specimens of the male of this most beautiful butterfly have a pinkish bloom on the orange area of the upperside. The female is yellower, with a white patch at the tornal margin of the forewing above.

Eggs are laid, many at a time, on tendrils or young shoots of *Adenia* (Passifloraceae), and the young larvae are gregarious. These are wine-red, with an enamel-white saddle mark, making them very conspicuous when half a dozen or so feed together in a well 'dressed' rank. They are said to be poisonous; if so, the butterfly may be too.

Similar, but less common, is *C. penthesilea*; it is distinguished from *hypsea* by a thin white submarginal line on the underside. A third species, a hill insect, and the most beautiful of the three, is *C. biblis*. With no broad white area near the apex of the forewing above, and an underside more finely and regularly flecked, it is at once distinguishable. A rare form of the female of the last has the upperside orange replaced by dull green.

THE PEACOCK PANSY (*Precis almana javana*) Plate 13H ♂

The 'Pansies' are common and gay little garden and roadside butterflies; of them all, this is the most familiar and widespread. The larva is said to feed on *Mimosa pudica*, the Sensitive Plant.

The BLUE PANSY, *P. orithya wallacei*, illustrated on Plate 13G, is the prettiest. It is locally common on open lawns, golf-courses and similar places on the plains, where the close mowing of the grass permits the growth of *Striga*, a tiny parasite on grass roots. On the tips of this plant the eggs are laid; and the larvae, a few days after hatching, appear to become night feeders, hiding amongst grass roots during the day and wandering in search of fresh food in the evening.

The female bears more prominent ocelli on the hindwing above, but sometimes has little or no blue colouring.

The GREY PANSY, *P. atlites*, a paler and duller insect than either *almana* or *orithya*, with smaller and more numerous ocelli, is also very common. Another fairly plentiful insect is the dark greyish brown *P. iphita*, which is replaced in Singapore by the redder brown *P. hedonia*.

THE GREAT EGG-FLY (*Hypolimnas bolina bolina*) Plate 14B ♂, C ♀

This insect is widely distributed over South and South-east Asia, but its history in Malaya is odd. Corbet records that the insect was rare in Malaya at the end of the nineteenth century and finally became extinct. The females of that race were black, with white submarginal spots. After thirty years of absence at the beginning of this century, the species again appeared and quickly established itself in open country on the Malayan plains; but this is a different race, the race found in Java.

The female is extremely variable. In Singapore specimens are taken without a trace of blue, and so pale a brown as to be considered mimics of *Danaus chrysippus*, and others, at the other end of the scale, without a trace of brown, and black enough to suggest that they have

PLATE 13. A. Rustic; B. Vagrant; C. Banded Yeoman; D. Royal Assyrian; E. Cruiser ♂, F. ♀; G. Blue Pansy ♂; H. Peacock Pansy ♀.

PLATE 13

PLATE 14

descended from the old *incommoda* race which flew in the peninsula sixty years ago. But the majority resemble the one illustrated.

Observations made elsewhere suggest that variations are seasonal. This does not seem true of Malaya; but more data are needed, particularly of the results of breeding the insect under different conditions.

A commoner but less interesting insect is *H. antilope*, which is black, sometimes washed with blue and with two series of submarginal spots. Slower than *bolina* in flight, *antilope* is a reasonable mimic of one of the 'Crows'. It can be distinguished from the black forms of *bolina* by the much narrower wings and less crenate margins.

THE AUTUMN LEAF (*Doleschallia bisaltide pratipa*) Plate 14E

This insect, common on forest roads, rocks and quarries, will be recognized from the plate. The underside is veined and mottled so as to suggest a partly decayed dead leaf, but it commonly settles with wings spread and flattened in the sunlight. The larva is said to feed on the leaves of jak fruit.

Related to *Doleschallia*, but rare in Malaya, is the much larger, true LEAF BUTTERFLY, *Kallima paralekta*; the upperside is a deep blue and orange.

THE STRAIGHT-LINE MAPWING (*Cyrestis nivea nivalis*) Plate 14F

This, the commonest 'Mapwing', is found in favoured spots at low altitudes. Its gliding flight is punctuated by rapid wingbeats and sudden turns; and it frequently alights on rocks or roads with wings outspread flat. The females are shyer and prefer the shade.

Above 2,000 feet, *nivea* is replaced by *C. maenalis*, which is darker, but otherwise similar in appearance and habits.

Related to the Mapwings, are the tiny yellow-brown MAPLET butterflies of the genus *Chersonesia*, which love dense forest and settle on the underside of leaves.

THE BURMESE LASCAR (*Neptis heliodore dorelia*) Plate 15A

Eighteen species of *Neptis* are known from Malaya; two of the commoner are illustrated, and two or three can be described fairly satisfactorily. For the rest the student must go to the key in Corbet and Pendlebury's book, or to one of the Museum collections.

PLATE 14. A. Malay Lacewing ♂, D. underside; B. Great Egg-Fly ♂, C. ♀; E. Autumn Leaf; F. Straight-Line Mapwing.

N. heliodore is one of three common species with uppersides black and yellowish; the three can be distinguished as follows:

Underside with solid black markings corresponding more or less with the upperside markings; forewing vein 10 arising from cell, not from vein 7 .. .. *N. heliodore*

Underside markings mostly only *outlined* with black, and black-dusted between the outlines .. .. .. *N. paraka*

Underside markings brown, broken and mottled .. *N. hordonia*

The larva of *N. heliodore* feeds on dead leaves of *Cratoxylon*, taking six or seven weeks to mature, a long time for a small butterfly. The young larva hides between two flaps of leaf which it bites off and secures to the leaf-vein. Later it bites through the stalk, securing the leaf or spray with silk, so that the leaves hang down and fade. Amidst these it is well camouflaged, and indeed partly hidden, and on them it feeds.

The Common Sailor (*Neptis hylas marmaja*)

Plate 15B upperside, C underside

This is the commonest of those *Neptis* species which are black and white above, and its yellow underside is distinctive. It is common on forest edges and often in parks and gardens. The larval food-plant is said to be Straits Rhododendron, *Melastoma*.

In the forests the commonest species is *N. nata*; it is a similar but slightly darker species above, and the underside is grey, not yellow. The larva of *nata* feeds on *Gironniera* spp., and closely resembles the hairy buds of those trees.

One other species of *Neptis*, *N. columella*, is common, particularly near towns. It is larger than *hylas*, the underside brown rather than yellow, and the white spots are not edged with dark brown, as in *hylas*. On freshly caught males, moreover, the upperside white seems to have a faint blue-green gloss. The larva feeds on *Cratoxylon*, but seems to prefer *Pterocarpus* (Angsana); like that of *N. heliodore* it makes a tent of dead foliage, in which it hides, eats and pupates.

The Lance Sergeant (*Parathyma pravara helma*) Plate 15D

Despite the pointed forewings and more powerful flight, *Parathyma* species are closely related to *Neptis*. From several somewhat similar insects, *P. pravara* is easily distinguished by the unbroken white cell streak on the forewing: i.e. it is not divided into a basal streak and a round or triangular spot, nor is it in any degree broken or constricted. It is common on the mainland at altitudes up to about 4,000 feet, but does not seem to occur in the Langkawi Islands or Singapore.

P. perius, sometimes common on the mainland, always rare in Singapore, recalls the colouring of *Neptis hylas*, having a yellowish underside, with the borders and most of the white spots outlined with black. Above, the cell streak is divided completely, not by mere constrictions, into four: a short quarter-inch basal streak, and three spots increasing in size towards the apex. *P. reta moorei*, another fairly common insect, has cell markings not very different from *P. perius*, but the basal streak is always joined to the nearest spot, and this is nearly always joined to the second; only the largest, most distal, spot is quite separate, and this is more triangular than in *perius*. The underside is quite different, being brownish grey. *P. kanwa* is another common species with a brownish-grey underside; above, the forewing cell markings are divided into a long (half-inch) basal streak, and a triangular spot.

The young larvae of most if not all *Parathyma* species are brown with short, knobbly branched spines. They eat away the leaf on each side of the central vein, and on this bare vein they rest. They collect all the particles of their own excrement and spin these into a roll or mass which resembles themselves, making, in short, inedible dummy caterpillars to discourage insectivores. The adult larvae vary from species to species, and so do the pupae. The adult larva of the last-named, *P. kanwa*, is brown with delicate green branched spines, and resembles a growth of moss on a decaying patch of leaf. Its pupa, made up of a series of silvery reflecting plates, outlined with black, resembles a pupa case which, since one seems to see through it, is empty. The food-plant is *Uncaria gambier*.

THE COLOUR SERGEANT (*Parathyma nefte subrata*)

Plate 15G ♂, H ♀-f. *neftina*, I ♀-f. *subrata*

All the *Parathyma* are forest insects, but *P. nefte* is sometimes seen in secondary growth, parkland and the like. The males love the morning sunlight, in which their markings seem a more dazzling white than those of their congeners. Later in the day they are less often seen than the females, and perhaps retire into deeper forest, or perhaps to the tree tops. The dark *subrata* form of the female is much less common than the other.

The young larva, like those already described, spins up its pellets of excrement together with bitten-off bits of leaf to make a dummy to delude insectivorous enemies. When half-grown, it develops three faded green dorsal patches in imitation of bits of leaf which, in the dummy, may not have faded. The adult caterpillar is green, with wine-red spines; the usual food-plant is *Glochidion*.

THE COMMANDER (*Moduza procris milonia*) Plate 15E

This active butterfly is less local in its habits than its relations of the genus *Parathyma*, many of which seem to have colonies and favourite glades and clearings and, although strong in flight, to return repeatedly to the same sunny sprays. *M. procris* is common throughout the plains, frequenting both primary and secondary forest, and coming also into the greener suburban areas. It has an air of hurrying off: along the road, over the hedge, up and over some high ridge of trees if it cannot find a way through after a couple of sallies.

In its early stages the larva is like those of the *Parathyma* species already described, but is more of an artist in its sculpturing of dummies, not being satisfied with a shapeless mass, but making two distinct rolls, which it pushes down the leaf as it feeds. It remains brown throughout, but in its last instar wanders far from its dummies, relying for disguise upon twisted postures which break its symmetry. The pupa is like a small withered leaf. Food-plants include *Timonias*, *Nauclea*, *Uncaria* and other plants of secondary growth and jungle edges.

Over about 4,000 feet are found two closely related[1] butterflies, *Limenitis daraxa* and *L. agneya*. Both are smaller than *M. procris*, and in so far as they live in small colonies, flying repeatedly between a few favoured basking places, they differ in habits. But in appearance they are somewhat similar, being dark brown above, with a band of pale spots (greenish white; not white as in *M. procris*) across both wings. At the apex, *L. daraxa* has three separated spots in line with

[1] By some authorities *M. procris* has been included in the genus *Limenitis*.

PLATE 15. A. Burmese Lancer; B. Common Sailor upperside, C. underside; D. Lance Sergeant; E. Commander; F. Clipper; G. Colour Sergeant ♂, H. ♀-form *neftina*, I. ♀-form *subrata*.

PLATE 15

PLATE 16

the rest, whereas *L. agneya* has a single small apical spot, in line with the main band, but three others almost at right angles.

THE CLIPPER (*Parthenos sylvia lilacinus*) Plate 15F

Except in the extreme north of Malaya, where it is fairly common, this lovely butterfly is local and rather rare. Its size makes it conspicuous, however, and it is to be found in most collections. It has a rapid gliding flight, marked by sharp turns and occasional quick wing beats. Corbet describes it as visiting *Lantana* flowers and frequenting rather open country in the north, and as having a lofty flight. I have seen it only in one locality, but on several occasions, in the fresh-water swamp jungle of S.E. Johore, darting and gliding low over a river. It is recorded from Singapore, but may not still survive there. *Adenia* and *Tinospora* are given as food-plants.

THE MALAY VISCOUNT (*Tanaecia pelea pelea*) Plate 16D ♀

In Singapore this is the commonest jungle butterfly, found wherever there is an acre or so of woodland, at most times of the day, and in almost all weathers. Through the rest of Malaya it is not so abundant. The sexes are alike on the upperside; below, the male is distinguished by jet-black V- and U-shaped markings.

T. pelea has been bred once, from an egg found on *Ardisia colorata*. Doubtless so common an insect feeds on other plants. The larva is similar to the striking larvae of the allied genus *Euthalia*: it has a row of long spines on each side of the back which project horizontally and also branch horizontally, so that the larva appears to be some kind of web. To turn, it must erect the spines on the inside of the curve. It is grey-green with purplish-grey oval dorsal markings.

HORSFIELD'S BARON (*Euthalia iapis puseda*) Plate 16F ♂, G ♀

This, the commonest species of a large and important genus, is found in open forests and forest edges, and also in parks and gardens. The females of this insect (as may be seen from the illustration) and those of a number of related species, are all superficially similar to *Tanaecia pelea*.

Another common butterfly often found in gardens and near towns and villages is the BARON, *E. aconthea gurda*, the male of which is

PLATE 16. A. Archduke ♂, B. ♀; C. Common Nawab; D. Malay Viscount ♀; E. Baron ♂; F. Horsfield's Baron ♂, G. ♀.

shown on Plate 16E. The larva feeds on cashew and mango and also on kinds of mistletoe. The adults are fond of rotting pineapple.

The ARCHDUKE, *E. dirtea dirteana* (Plate 16A ♂, B ♀), is one of the largest and handsomest members of the genus, and also one of the commonest. It is essentially a jungle butterfly, and is often disturbed on shady jungle paths where it settles with outspread wings to feed on fallen fruit. A much rarer species, both sexes of which closely resemble the female of *E. dirtea* on the upperside, is *E. canescens*. The undersides are distinct: *canescens* being yellowish, and the female of *dirtea* being washed with blue on the hindwing. Occasionally the two are found together, but *canescens* usually prefers drier forest where it basks on sheltered sprays in patches of sunlight. It is worth while trying to catch any butterfly which seems to be a rather small female *dirtea*, particularly if it seems yellowish in flight. Both species are swift on the wing, and need careful stalking.

On the hills *E. lepidea* is fairly common: both sexes remind one of *iapis* males, but the blue wing borders are replaced by greyish buff.

THE COURTESAN (*Idrusia nyctelius euploeoides*) Plate 2H ♂, I ♀

The females of this butterfly occur in two mimetic forms, one of which, ♀-f. *euploeoides*, a mimic of the female of *Euploea diocletianus*, is shown alongside that butterfly on Plate 2. The other, ♀-f. *isina*, mimics the male of that butterfly. The insect is widely distributed, and the males, though not very common, are much commoner than the females.

THE COMMON NAWAB (*Polyura athamas samatha*) Plate 16C

The 'Nawabs' are all large handsome insects, more beautifully, though cryptically, marked on the underside. *P. athamas* is the most widely distributed, and the commonest except in Singapore, where it is rare, and where the species *P. hebe*, apparently uncommon elsewhere in Malaya, is abundant.

The larva of *P. hebe* is green, the head with an elaborate four-horned mask, and the body thick in the middle and tapering fore and aft; each segment is scalloped with darker green so that on the leaflets of *Albizzia* or *Adenanthera*, on which it feeds, it is beautifully camouflaged. The pupa is silvery green, so smoothly rounded that wing-cases and segmentation are almost invisible, and distinctly stalked, so that the likeness to a large unripe berry is complete.

The larva of *P. athamas* is less elaborately marked, and the pupa less rounded and less stalked. The food-plants given are *Albizzia*, *Grewia*, *Caesalpinia* and *Acacia*; Davidson and Aitken give *Delonix* (Flame of the Forest) as an additional food-plant in India.

Closely related to *Polyura* is the genus *Charaxes*: butterflies with stout bodies, similar forceful-looking forewings, but more compact hindwings. They have an aggressive flight, returning repeatedly to the same lofty branch commanding a forest glade; and they are strongly attracted to dung and rotting fruit. Only one, the TAWNY RAJAH, *C. polyxena*, is not rare; it is shown on Plate 17.

RIODINIDAE THE RIODINIDS

Smallish butterflies, rather uniform in size, and fairly similar to one another in general appearance and habits (Plate 17, B, C, D, K, L and M). Most are brown, several with a purplish flush; the pretty *Laxita* species are carmine, an unusual colour in butterflies. All frequent the forest on sunny days, and settle restively with wings half closed. Little is known of the early stages of Malayan species.

THE PUNCHINELLO (*Zemeros flegyas albipunctata*) Plate 17D ♀

A gay, active little butterfly, often to be seen until late in the afternoon on sunny days. The males are darker and less distinctly marked.

THE MALAYAN PLUM JUDY (*Abisara saturata kausambioides*) Plate 17K ♂, L ♀

This is but one of several fairly similar little butterflies, all denizens of the forest, and all with a habit of settling with half-open wings, and moving restlessly about on a sunlit leaf.

Two species of the genus, *A. neophron* and *A. savitri* are quite distinct: both have tails instead of the mere projection at hindwing vein 4. *A. neophron* occurs on the hills: it is blackish brown and has a clear-cut white bar from mid-costa to the tornal angle of the forewing. *A. savitri* is pale greyish brown, with two indistinct paler grey bands on the forewing. The latter is common in Singapore, but rare elsewhere in Malaya.

THE MALAY RED HARLEQUIN (*Laxita damajanti*) Plate 17M ♂

This, one of the prettiest of Malayan butterflies, occurs not uncommonly at moderate elevations on the hills. The female is paler than the male. The underside is beautifully marked with black streaks crossed with metallic blue. It frequents dense forests, and its habits are similar to those of other Riodinids.

THE SMALL HARLEQUIN (*Taxila thuisto thuisto*) Plate 17B ♂, C ♀

That the two *Taxila* species are related to *Laxita* is indicated by the anagram; they lack the brilliant reddish-purple colour, but they are quite handsome insects. The upperside of the female of *T. thuisto* gives some indication of the underside pattern of both sexes, but the underside ground colour is brighter, the black streaks more sharply defined, and the cross-streaks (shown as grey in the illustration of the upperside) are silvery blue, more brilliant still in the male.

T. haquinas is a slightly larger and duller insect: the male brownish black with a dull brown forewing area; the female with a diffuse pale apical area instead of the sharp-defined spots of *thuisto*. The underside also lacks the metallic brilliance which makes *T. thuisto* and *L. damajanti* such attractive insects. Unlike *L. damajanti*, the two *Taxila* species seem to prefer the plains, and both are occasionally found in Singapore.

LYCAENIDAE THE BLUES AND HAIRSTREAKS

With a few exceptions these are smallish, sometimes very small butterflies; many are brilliantly coloured with metallic blue, silvery or coppery scales (Plate 17, figs. E to J and Plates 18 and 19). The hindwings of some species have one, two or sometimes even three pairs of long filamentous 'tails', those with one or two pairs usually bearing underside eyespots on a lobe of the wing. There is evidence to suggest that these spots (together with the thin tails which, moved by a slight breeze, resemble the butterflies' antennae) delude an enemy into thinking the back is the front, and divert a peck from a vital to an unessential part of the butterfly.

The eggs are variable; often flattened. A few species (particularly in the genera *Liphyra*, *Miletus*, *Spalgis*) have larvae which eat not leaves, but other insects. Many larvae are wood-louse shaped with heads, legs and claspers hidden under a fleshy carapace; some are attended by ants, and in return for the ants' protection supply a sweet liquid from a dorsal gland. The pupae vary in form and in means of attachment. A few are said to pupate unattached amidst dead leaves; some are held upright by a girdle; some pupate in rolled leaves lined with silk (virtually a cocoon); some are without a girdle and are attached thickly and rigidly by the anal segment, and curve down from an upright support.

THE SUMATRAN GEM (*Poritia sumatrae sumatrae*) Plate 17I ♂, J ♀

The males of this and allied genera (sub-family Poritiinae) are gay little butterflies usually with brilliant metallic scales contrasted with black. In this species the female, though less bright, is almost as attractive as the male. The underside is crossed by a series of red-brown parallel broken curved bands.

P. philota has a similar, but greyer underside. Above, the female is dull brown, sometimes with paler discal areas; and the male is black, spotted and streaked with metallic blue (in contrast with *sumatrae*'s broader greener areas).

P. sumatrae and *philota* are less uncommon than most Poritiinae. They may be found singly in shady forest on sunny days; but in the early mornings on a sunlit forest edge, they fly actively together, the males chasing one another, and the two sexes spiralling up to fifty or sixty feet in, presumably, prenuptial flight. Of neither species is the life-history known.

BIGGS'S BROWNIE (*Miletus biggsii biggsii*) Plate 17G

Colonies of these rather shadowy little butterflies, with their rising and falling flight, remain round certain forest trees or bushes sometimes for years. The one species of which the life-history is known feeds on aphids, and perhaps the stability of these insects accounts for that of the butterfly colonies.

M. biggsii is one of the commoner 'Brownies', and may be recognized by the comparatively narrow, fairly elliptical white patch on the forewing. *M. symethus* has larger pale areas, and frequents somewhat more open country.

THE LESSER DARKIE (*Allotinus unicolor*)
Plate 17E ♂, F ♀ underside (subsp. *unicolor* from Singapore)

The species of *Allotinus* have the underside whitish speckled with darker colour, whereas *Miletus* spp. are scalloped brown beneath. Most *Allotinus* males have a long elliptical brand on vein 4 of the forewing above.

The species are local, but *A. unicolor* is sometimes plentiful where it occurs. The race found on the mainland, *A. unicolor dilutus*, has a darker underside than those illustrated.

THE STRAIGHT PIERROT (*Castalius roxus pothus*) Plate 18G underside

These little butterflies flutter zigzaggedly close to the ground, the black and white flickering uncertainly. They are fond of settling on damp sand. *C. roxus* is common in Malaya, but is not found in Singapore, where the somewhat similar *C. elna* is abundant.

THE COMMON HEDGE BLUE (*Celastrina puspa lambi*)
Plate 18A ♂, B underside

This common butterfly frequents forest edges and roadsides on the plains. The female has the wing centres lustrous pale greenish blue, and broad black borders. On the hills are found a score of different *Celastrina* species, some identifiable only with difficulty.

THE LESSER GRASS BLUE (*Zizina otis lampa*) Plate 18C ♂, D underside.

This tiny butterfly is abundant on every scrap of grassland on the plains. The female is blackish brown, usually with some blue scaling near the wing bases. The larva is said to feed on the Sensitive Plant.

Two other tiny Grass Blues are found in Malaya. Both have whitish (not pale buff) undersides, and the spots more distinct. *Zizeeria knysna* is sporadically common but normally scarce; it is a deeper blue than *otis*, and can be recognized by a spot in the forewing cell beneath. *Zizula hylax*, the smallest Malayan butterfly, with a wing expanse of sometimes only 13 mm., has no spot in the cell, but a distinctive one in space 11 on the forewing costa. *Z. hylax* is locally common near the coast.

THE PEA BLUE (*Lampides boeticus*) Plate 18E ♀, F underside

This is the 'Long-tailed Blue' which occasionally reaches England. The underside figure will easily identify it. The larvae feed on the seeds of various peas and beans and, in Malaya, seem to favour *Crotalaria*, which is used as a cover on rubber plantations. The insect is found almost everywhere on the plains, and is common also on some hill stations.

THE SKY BLUE (*Jamides caeruleus caeruleus*)

Plate 18H ♂, I ♀ underside

This is a deeper brighter blue than most of the other 'Ceruleans'; and the female too is distinctly blue. The species is common.

Two common species, *J. celeno* and *J. pura* are very pale. Males of the former lack metallic sheen and have broad black forewing borders (more than 1 mm.); those of *pura* have thin borders, pointed apices, and a distinct pale metallic sheen. The females are more alike, but *celeno* is slightly duller. Both are found in the forest; *celeno* occurs also in open country. Another distinctive insect is *J. bochus*: the underside is grey-brown with the white lines very thin. Above, the black border is exceptionally broad in the male, making the small area of blue seem specially brilliant; the female is distinctly blue, though chalky. *J. bochus* prefers open country and shore vegetation; it can always be found on *Andira* blossom and, on the shore, the females are sometimes seen ovipositing on the flowers of *Derris uliginosa*.

THE COMMON LINE BLUE (*Nacaduba nora superdates*)

Plate 17H underside

The Line Blues are closely related to the Ceruleans, but the males are dull blue or purple above. Several of the eighteen species found in Malaya are plentiful, but few can be identified certainly and easily, and the collector is advised to collect and set a large number, sort them into series, and then work through the key in Corbet and Pendlebury's book. *N. nora*, perhaps the commonest species, is also easy to identify. The male is pale mauve, and the female blackish brown above; the rounded wings and greyish-buff underside are helpful in identifying the species.

THE MALAYAN SUNBEAM (*Curetis santana malayica*)

Plate 18O ♂, P underside

The Sunbeams, copper-red above and silver-grey beneath, are attractive and, as a group, distinctive; but the species are not easy to identify. The form of the underside post-discal bands (in *santana* broad and black-dusted and, on the forewing, slanted towards the apex) is one aid to identification; another is the upperside distribution of black, which, in *santana* males, is continued from the forewing termen some way along the dorsal margin. In the females the black is more extensive, and the orange paler. On the female of the somewhat similar *C. bulis*, the orange is replaced by white.

Curetis larvae feed on leaves and young climbing shoots of *Derris* and similar plants. They are pinkish green with oblique lateral markings closely resembling a young unexpanded pinnate leaf. On the last segment but one are two erect fleshy tubes from which the caterpillar when alarmed can thrust flail-like organs and whirl them about. These are modifications of scent-glands which in some Lycaenids attract ants to honey secretions. The green pupa is the shape of an egg halved by slitting lengthwise; it is secured to a leaf by a girdle.

THE CENTAUR OAK BLUE (*Narathura centaurus centaurus*)

Plate 19A ♂, J underside

There are over eighty species of Oak Blues (all of which were once included in the single genus *Arhopala*), so we must be content with describing only a few of these beautiful insects. *N. centaurus*, one of the largest, is perhaps the commonest, and is found not only in the jungle, but also around trees in towns (e.g. in front of Raffles Museum, in the heart of Singapore). The greenish outlines of the forewing discal spots below will always identify it. The female, like all those of the group, has broad black wing-borders, and is usually a lighter blue.

Another large species, *N. aedias agnis*, is metallic purple above; on the hair-brown underside the spots are rounded and separated. The small tailless *antimuta* (Plate 19D ♂, E ♀) is common; the faintness of the underside markings, and the male's very dark velvety blue and narrow black borders, will help to identify it. Another

PLATE 17. A. Tawny Rajah; B. Small Harlequin ♂, C. ♀; D. Punchinello ♀; E. Lesser Darkie ♂, F. ♀ underside; G. Biggs's Brownie; H. Common Line Blue; I. Sumatran Gem ♂, J. ♀; K. Malayan Plum Judy ♂, L. ♀; M. Malay Red Harlequin ♂.

PLATE 17

PLATE 18

common tailless species, but larger, has a distinctive oval patch of modified scales in the centre of purple-blue forewings in the male; this is *N. epimuta*. The males of a few species are metallic green, the commonest being *N. eumolphus maxwelli*.

The green species illustrated on Plate 19 (B ♂, C ♀) is *Aurea trogon*, of which a close relation, *A. aurea*, differs in having the brown forewing border increase in width to about 1 mm. at the tornus. *A. aurea* may further be distinguished by its brown underside; that of *trogon* being washed with deep purple. Both may be distinguished from the green species of the genus *Narathura* by their short thick tails.

Most butterflies of the group frequent lowland forest and are shy and elusive. The larvae feed on *Quercus* (oak), *Terminalia*, *Eugenia*, *Lagerströmia*, *Macaranga*, etc., and are guarded by ants (the Karingga, *Oecophylla smaragdina*, tends *centaurus* larvae) in return for a sweet secretion which the ants 'milk' from a dorsal gland. The larva sometimes (not always) pupates in a rolled sewn leaf where ants continue to protect it until the insect emerges.

THE YAM FLY (*Loxura atymnus fuconius*) Plate 18J

This insect flies weakly in open forest and secondary growth. The larvae closely resemble the polished green fleshy shoots of *Smilax* and *Dioscorea* (yam) on which they feed. Both *atymnus* and the darker *L. cassiopeia* frequent the plains; on the hills occurs a superficially similar insect, but darker still and with the hindwing dark bordered and branded, and with the forewing termen curved: *Yasoda pita*. Again *Smilax* is the food-plant.

Another butterfly, differently coloured but with similar habits and life-history, is the BRANDED IMPERIAL, *Eooxylides tharis*. The underside is shown on Plate 19M; the upperside is black, with a circular brand of modified scales in the centre of the forewing of the male, and the hindwing tornal area marked with white and black much as on the underside. In Singapore, where it is abundant, it may be seen settled on the climbing shoots of *Smilax* and fluttering about the forest edges. On the mainland it appears to be local. The FLUFFY TIT, *Zeltus amasa maximinianus* (Plate 19K ♂, L underside) is common along paths and at forest edges on the lowlands.

PLATE 18. A. Common Hedge Blue upperside, B. underside; C. Lesser Grass Blue ♂, D. underside; E. Pea Blue ♀ upperside, F. underside; G. Straight Pierrot underside; H. Sky Blue ♂, I. ♀ underside; J. Yam fly; K. Common Posy ♂, L. underside; M. Peacock Royal ♂, N. ♀; O. Malayan Sunbeam ♂, P. underside; Q. Black-banded Silverline underside.

THE COMMON POSY (*Marmessus ravindra moorei*)

Plate 18K ♂, L ♀ underside

Unlike the long-tailed butterflies mentioned in the preceding section, this common and charming little forest insect has a swift darting flight. The female lacks the metallic blue scaling on the upperside, and usually has a yellowish patch in the discal area of the forewing. The food-plants of the pink and grey larva include *Derris scandens*.

THE PEACOCK ROYAL (*Pratapa cippus maxentius*) Plate 18M ♂, N ♀

The *Pratapa* species are all brilliant blue or greenish blue with pale, usually grey, undersides. The larvae feed on mistletoe (*Loranthus*), and are easy to breed, but difficult to find, for they resemble curls of leaf, stalk ends, and other parts of the plant. The butterflies fly rapidly and favour hill-tops and lofty branches where *Loranthus* is growing. They seem to prefer morning and afternoon sunlight to that of midday. *P. cippus* is by far the commonest species.

THE BLACK-BANDED SILVERLINE (*Spindasis syama terana*)

Plate 18Q underside

The two species of *Spindasis* are easy to recognize: *S. syama* has the silver lines edged with black; *S. lohita*, the commoner, with red. Above, the wings are black, shot with blue in the males, with a yellow tornal area. Both species frequent flowering shrubs, particularly *Cordia*, in open grassland.

THE COMMON RED FLASH (*Rapala iarbus iarbus*) Plate 19I ♂

The species of *Rapala* are active around trees and flowering shrubs (*Eugenia, Cordia, Clerodendron*) in morning and late afternoon sunlight. *R. iarbus* can be recognized from the illustration; the female is a paler, more reddish-brown insect than G on the same plate, and the underside of *iarbus* is grey. Figures F, G and H on Plate 19 represent the male, female and underside respectively of another common species, *R. dieneces*; but several others are not unlike, so the collector should check that the underside is brownish ochre, and the male brush on the underside of the forewing, at mid-dorsum, is dark brown. Fairly common also is *R. pheretima*, recognizable by the broad cell-end bar on a hair-brown underside; above, the male is coppery red, the female steely blue.

HESPERIIDAE The Skippers

It has several times been suggested that the Skippers are so distinct that they should be grouped neither with butterflies nor with moths. Structurally they differ from other butterflies in that all the wing veins arise either from the base of the wing or from the cell. Plate 20 shows only Hesperiids, and it will be seen that many are larger insects than they first appear: many butterflies with much smaller bodies but large wings, *look* larger. As we should expect from the small wings, most typical Skippers are rapid fliers. Many love sunshine, and may be seen darting around *Lantana* and *Cordia* bushes; but most, perhaps, are active only in the early morning and after about 5.30 in the evening, although such species may be found nearer the middle of the day flying in shadier parts of the forest.

The larvae of many species feed on monocotyledons: palms, grasses, bamboos, gingers, bananas. Most feed and pupate sheltered in a roll of leaf where protective colouring is pointless, so many are white. Pupae are usually attached by hooks and pad, and sometimes supported by more than one girdle; occasionally the proboscis sheath projects beyond the last abdominal segment.

The Orange Awlet (*Bibasis harisa consobrina*) Plate 20B ♂, C ♀

In this common species the female has the brighter colouring. Both sexes fly powerfully with a clicking noise at dusk and also in the early morning, visiting flowering trees and shrubs; it is especially fond of *Eugenia longiflora*. The larva is white, yellow and olive-green, spotted and reticulated with black, and has a black-spotted orange head. It feeds in the rolled leaves of *Arthrophyllum diversifolium.*

The Plain Banded Awl (*Hasora vitta vitta*) Plate 20D ♀ underside

The English names of this and the preceding species seem inexplicable. This insect and its congeners fly swiftly at dusk and dawn, and sometimes, in shady parts of the forest, in the daytime. Some are fond of caves and moist rocks. The upperside of *H. vitta* is plain brown in the male, with a single pale apical spot; the female has two larger spots in addition. *H. badra*, with a single white spot in the cell on the underside of the hindwing, instead of the broad white line which characterizes *vitta*, is almost as common. Fairly common on the hills, and much more conspicuous, is *H. schoenherri*, which has a broad bright yellow band on the hindwing above and below. All three insects occur in Singapore, but the last is very local there.

THE DARK YELLOW-BANDED FLAT (*Celaenorrhinus aurivittata cameroni*) Plate 20N

This distinctive insect is not uncommon on forest roads. It flies in a more leisurely way than some related species, and settles with wings flattened.

The SNOW FLATS, of which *Tagiades gana gana* (Plate 20E) is one of the commonest, fly more swiftly, their white areas producing the illusion that they are much smaller entirely white insects. The Flats are forest insects, and fly by day, not at dusk.

THE BUSH HOPPER (*Ampittia dioscorides camertes*) Plate 20F ♂, H ♀

This little butterfly, both sexes of which are figured, may be seen visiting flowers on open grassland at the forest edge, and may easily be distinguished from the many Skippers of rather similar colouring which abound at flowers and flowering shrubs in such situations.

Of these may be mentioned the LESSER DART, *Potanthus omaha omaha* (Plate 20I ♂), one of many very similar species, and the PALM DART, *Telicota augias augias* (Plate 20G ♂), a larger insect, and again one of a group. *T. augias* is separable from some of its congeners by the more *orange* shade of its yellow; and another very common species, *T. colon*, is fairly easy to identify by the fact that the yellow colour of the forewing band is continued narrowly along the veins almost to the termen. The females both of *Telicota* and *Potanthus* species are darker than the males.

The YELLOW-VEIN LANCER, *Plastingia latoia latoia* (Plate 20K) is far less common, but may sometimes be taken at flowering shrubs or trees in the morning sunlight. It is more gaily coloured than most Skippers, the veins below being picked out in bright yellow.

PLATE 19. A. Centaur Oak Blue ♂, J. underside; B. *Aurea trogon* ♂, C. ♀; D. Small Tail-less Oak Blue ♂, E. ♀; F. *Rapala dieneces* ♂, G. ♀, H. underside; I. Common Red Flash ♂; K. Fluffy Tit ♂, L. underside; M. Branded Imperial underside.

PLATE 19

PLATE 20

THE CHESTNUT BOB (*Iambrix salsala salsala*) Plate 20Q ♂

The pale post-discal spots on the upperside of the forewing distinguish males of *salsala* from other species. But the underside too is distinctive: the hindwing and costal half of the forewing are thickly dusted with yellow; the hindwing lacks a spot in the cell, but just beyond the cell, in the centre of the wing, is a conspicuous one, encircled with black. There are other silvery spots, but they are hardly noticed, since the central spot dominates the wing.

I. stellifer, the STARRY BOB (Plate 20J underside) is not quite so common as *salsala*, and occurs less often outside the forests. The underside is darker, more silvery spots are noticeable, and in particular there is a small spot in the hindwing cell, near the cell-end, and another in space 5, not near the cell-end as in *salsala*, but half-way between the cell spot and the termen.

THE GRASS DEMON (*Udaspes folus*) Plate 20O

This distinctive insect is widely distributed, but never abundant: single specimens are taken in gardens and open country.

THE BANANA SKIPPER (*Erionota thrax thrax*) Plate 20M

This and another common species, *E. torus*, which has a more rounded apex and a convex termen, are the 'Banana Rollers', the larvae and pupae of which can be found in rolled banana leaves.

An even larger Skipper, and darker, particularly on the underside, is the GIANT RED-EYE, *Gangara thyrsis*. It cannot be confused with *E. thrax*, because the large spot near the costa on the forewing is larger than the one behind it.

Another butterfly of the same group is the COCONUT SKIPPER, *Hidari iravi* (plate 20P underside), the larva of which feeds on palms and bamboo. It is abundant everywhere in the lowlands, flying at dusk and frequently sheltering in balconies and porches.

A more attractive Skipper, *Unkana ambasa batara* (Plate 20A ♀) is not uncommon around town and village gardens, and is occasionally attracted after sundown by light. The male hindwing is unmarked above; in both sexes the hindwing below is broadly washed with chalky white, the veins remaining narrowly dark chocolate. The larva feeds in rolled-under pandan leaves.

PLATE 20. A. Hoary Palmer; B. Orange Awlet ♂, C. ♀; D. Plain Banded Awl underside; E. Large Snow Flat; F. Bush Hopper ♂, H. ♀; G. Palm Dart ♂; I. Lesser Dart ♂; J. Starry Bob underside; K. Yellow-vein Lancer; L. Small Branded Swift; M. Banana Skipper; N. Dark Yellow-banded Flat; O. Grass Demon; P. Coconut Skipper underside; Q. Chestnut Bob; R. Contiguous Swift.

THE CONTIGUOUS SWIFT (*Polytremis lubricans lubricans*) Plate 20R

This, and the SMALL BRANDED SWIFT, *Pelopidas mathias* (Plate 20L ♂) are abundant on flowers, especially *Cordia*, in gardens and grassland. The larvae of both feed on lalang.

There are so many related and similar species that again the beginner is advised to collect a large number, to divide them carefully into series and, with his eyes thus sharpened for significant differences, to consult Corbet's keys or one of the Museum collections.

INDEX

ENGLISH NAMES *are in ordinary type*
LATIN NAMES *are in italic type*
FAMILIES *are in italic capitals*

Numbers followed by letters are references to the Plates. Numbers not followed by letters are page references. Main descriptions are referred to in heavy type.

Note: In this book most of the butterflies are given their full names under the 'trinomial' system. The first name, with a capital initial letter, is the name of the GENUS to which the species belongs; the second is that of the SPECIES itself; and the third is that of the SUB-SPECIES or RACE. Normally, when we refer to a butterfly, only the generic and specific names are necessary; and only these are indexed. But the race names are reminders that a species found in Malaya may be markedly different in appearance or habits from the same species in other parts of its range; and, exceptionally, two or more races of a single species are found in Malaya; for example, there are three races of the Common Palmfly, *Elymnias hypermnestra*, viz. *E. hypermnestra agina*, *E. hypermnestra beatrice*, and *E. hypermnestra tinctoria*.